ESTIMATION OF MERCURY BIOACCUMULATION POTENTIAL FROM WASTEWATER TREATMENT PLANTS IN RECEIVING WATERS: PHASE I

by:

J. David Dean
ArcTellus

Robert P. Mason
University of Connecticut

2009

WERF

The Water Environment Research Foundation, a not-for-profit organization, funds and manages water quality research for its subscribers through a diverse public-private partnership between municipal utilities, corporations, academia, industry, and the federal government. WERF subscribers include municipal and regional water and wastewater utilities, industrial corporations, environmental engineering firms, and others that share a commitment to cost-effective water quality solutions. WERF is dedicated to advancing science and technology addressing water quality issues as they impact water resources, the atmosphere, the lands, and quality of life.

For more information, contact:
Water Environment Research Foundation
635 Slaters Lane, Suite 300
Alexandria, VA 22314-1177
Tel: (703) 684-2470
Fax: (703) 299-0742
www.werf.org
werf@werf.org

This report was co-published by the following organizations. For non-subscriber sales information, contact:

IWA Publishing
Alliance House, 12 Caxton Street
London SW1H 0QS, United Kingdom
Tel: +44 (0) 20 7654 5500
Fax: +44 (0) 20 7654 5555
www.iwapublishing.com
publications@iwap.co.uk

Library of Congress Catalog Card Number: 2008942452

IWAP ISBN: 978-1-84339-357-3/1-84339-357-3

This report was prepared by the organization(s) named below as an account of work sponsored by the Water Environment Research Foundation (WERF). Neither WERF, members of WERF, the organization(s) named below, nor any person acting on their behalf: (a) makes any warranty, express or implied, with respect to the use of any information, apparatus, method, or process disclosed in this report or that such use may not infringe on privately owned rights; or (b) assumes any liabilities with respect to the use of, or for damages resulting from the use of, any information, apparatus, method, or process disclosed in this report.

ArcTellus, University of Connecticut

The research on which this report is based was developed, in part, by the United States Environmental Protection Agency (EPA) through Cooperative Agreement No. CR83155901-1 with the Water Environment Research Foundation (WERF). However, the views expressed in this document are solely those of ArcTellus and the University of Connecticut and neither EPA nor WERF endorses any products or commercial services mentioned in this publication. This report is a publication of WERF, not EPA. Funds awarded under the Cooperative Agreement cited above were not used for editorial services, reproduction, printing, or distribution.

This document was reviewed by a panel of independent experts selected by WERF. Mention of trade names or commercial products does not constitute WERF, BACWA, nor EPA endorsement or recommendations for use. Similarly, omission of products or trade names indicates nothing concerning WERF's, BACWA's, or EPA's positions regarding product effectiveness or applicability.

ACKNOWLEDGMENTS

Report Preparation

Co-Principal Investigators:

J. David Dean
ArcTellus

Dr. Robert P. Mason
University of Connecticut

Project Team:

Van Buuren Consulting

Internal Peer Review Team:

Dr. Helen Hsu-Kim
Duke University

Tim Tuominen
Western Lake Superior Sanitation District

Dr. James Wiener
University of Wisconsin, LaCrosse, WI

Project Subcommittee

Michele Pla
Bay Area Clean Water Agencies (BACWA)

Dr. P.L. Fan, Ph.D.
American Dental Association

Keith Linn
Northeast Ohio Regional Sewer District

Richard Looker
San Francisco Bay Regional Water Control Board

Mike Connor, PSC Chair
San Francisco Estuary Institute

David W. Tucker
City of San Jose, CA

Vicky Fry
Sacramento Regional County Sanitation District

Water Environment Research Foundation Staff

Director of Research: Daniel M. Woltering, Ph.D.
Program Manager: Jane M. Casteline, M.S.

ASTRACT AND BENEFITS

Abstract:

This project was initiated in response to the establishment of mercury TMDLs around the country and issues raised by this process, specifically concerning the issue of mercury bioavailability. While many TMDLs recognize that point sources constitute a small fraction of the mercury load to a waterbody, a question has been raised concerning the *relative* bioavailability of mercury coming from various sources. For instance, is the mercury discharged from a wastewater treatment plant more or less bioavailable than mercury in precipitation, mercury in urban stormwater, or mercury in sediments? This project seeks to address this question by developing a reliable definition and approach to estimating bioavailability, by profiling various sources of mercury in a watershed with regard to the species of mercury present and by profiling those factors or conditions in either the effluent or the receiving water that enhance or mitigate the bioavailability of those forms.

There were several important objectives relevant to the estimation of bioavailability and potential bioaccumulation of mercury from wastewater treatment plants and other sources in receiving waters. The first was to develop a working definition of bioavailability. For purposes of this project, this definition includes not only methylmercury, the form of mercury that readily bioaccumulates in aquatic food chains, but also bioavailable and potentially bioavailable inorganic mercury species that can be converted to methylmercury. It is concluded that the strength of binding to solids and mercury-sulfur-organic matter associations are major factors in determining the bioavailability of inorganic Hg.

A second major objective was to identify those factors or conditions in both the effluent and the receiving waters that enhance or mitigate the transformation of inorganic mercury to methylmercury and its subsequent bioaccumulation. Profiles were developed for: 1) various sources of mercury in watersheds, including wastewater treatment plants, with regard to bioavailable and potentially bioavailable mercury; and 2) key factors in effluents and receiving waters that enhance or mitigate it. A procedure to assess the relative bioavailability of mercury from various watershed sources, including wastewater treatment plants, was developed and tested using data from a U.S. location. The project also features a literature review of conventional and emerging technologies for the removal of mercury from effluent streams and their effects on mercury bioavailability. A review of the salient aspects of mercury TMDLs completed by U.S. EPA and the states is also included.

It is important to note that this project deals mainly with the estimation of bioavailable species of mercury and how they are affected by wastewater treatment, although the acquisition of data to support bioaccumulation studies is addressed in the Phase II report. It is important to note that bioavailability does not necessarily equate to bioaccumulation. However, provided that bioavailable mercury species are present in an aquatic ecosystem, bioaccumulation will occur. The degree to which this happens will depend upon the biological species and the dynamics of the food chain present. Bioaccumulation and consumption must occur in order for humans or wildlife to be at risk.

This project concludes that, based on available data and bioavailability as defined in this report, wastewater effluent is one of the lowest among the sources evaluated with respect to mercury bioavailability due to its typically low levels of methylmercury. Due to their low levels of suspended solids and low levels of mercury in solids, wastewater treatment plants employing post-secondary treatment should not contribute appreciably to local sediment mercury burdens.

Benefits

- Provides a guidance document that allows wastewater treatment professionals to assess and manage their risk vis-a-vis mercury discharges, in light of their contribution to overall mercury loads in various types of receiving waters.

Keywords: Mercury, bioavailability, bioaccumulation, wastewater effluent, wastewater treatment, receiving waters, TMDLs

TABLE OF CONTENTS

LIST OF TABLES

LIST OF FIGURES

LIST OF ACRONYMS

ADA	American Dental Association
AVS	Acid volatile sulfides
BAF	Bioaccumulation factor
B-Hg	Bioavailable mercury
%B-Hg	Percent bioavailable mercury normalized to total mercury (also relative bioavailable Hg)
BNR	Biological nutrient removal
BOD	Biological oxygen demand
COD	Chemical oxygen demand
D-Hg	Dissolved phase mercury
D-MeHg	Dissolved methylmercury
DO	Dissolved oxygen
DOC	Dissolved organic carbon
DOE	Department of Energy
DOM	Dissolved organic matter
EMC	Event mean concentration
fB-Hg	Fraction of bioavailable mercury normalized to total mercury
FDA	Food and Drug Administration
GAC	Granular activated carbon
Hg	Mercury
Hg^0	Hg species in the nonionic inorganic form
Hg^{2+}	Hg species in the inorganic (2+) valence state
Hg^{II}	Hg species in the ionic inorganic (II) form
I-Hg	Inorganic mercury
IB-Hg	Inorganic bioavailable mercury
INR-Hg	Inorganic non-reactive mercury
K_D	Partition coefficient
LA	Load allocation
LIS	Long Island Sound
$\log K_D$	base 10 logarithm of the partition coefficient

LOI	Loss on ignition
MeHg	Methylmercury (equivalent to CH_3Hg^+ or CH_3Hg^{II})
%MeHg	Percent MeHg normalized to total mercury (also relative MeHg)
MGD	Million gallons per day
MOS	Margin of safety
NACWA	National Association of Clean Water Agencies
NOM	Natural organic matter
NPDES	National Pollutant Discharge Elimination System
OM	Organic matter
P-Hg	Particulate phase mercury
P-MeHg	Particulate methylmercury
POC	Particulate organic carbon
POM	Particulate organic matter
R-Hg	Reactive mercury
%R-Hg	Percent reactive mercury normalized to total mercury (also relative reactive Hg)
RO	Reverse Osmosis
SRB	Sulfate reducing bacteria
T-Hg	Total mercury
TMDL	Total maximum daily load
TOC	Total organic carbon
TSS	Total suspended solids
U.S. EPA	United States Environmental Protection Agency
UV	Ultraviolet
VSS	Volatile suspended solids
WLA	Waste load allocation
WQT	Water quality target
WWTP	Wastewater treatment plant

EXECUTIVE SUMMARY

Mercury levels in fish tissue and the associated human health concern have led regulatory agencies to issue fish consumption advisories for over 2000 waterbodies in the United States. The existence of fish consumption advisories in a waterbody have in turn been used as a basis for placing these waterbodies on the states' lists of impaired waters, the 303(d) lists. The sources of mercury are known to be diverse, as it is a naturally occurring element, although it is generally accepted by the scientific community that the emissions of mercury have increased several fold during the industrial period. Therefore, attempts to regulate mercury discharges to waterbodies have focused on point source dischargers and air emissions.

The TMDL process has been used to estimate mercury loadings to waterbodies – both to estimate the acceptable load that would result in the attainment of acceptable levels in fish tissue, and to estimate the current load of mercury to the waterbody that is resulting in nonattainment. In the case of mercury, a particularly difficult aspect of this exercise arises from the fact that it must be transformed from the form that is typically deposited or discharged to waterbodies (inorganic, primarily divalent mercury) to the form that is bioaccumulated in fish (methylmercury). Currently, there are no accepted techniques available to "trace" mercury from its emitted origin to its appearance in fish tissue.

The movement toward an assessment of the bioavailability of mercury from various mercury sources once it reaches a waterbody is a step in this direction. It is reasonable that a source contributing mercury which is more difficult to transform into a bioaccumulative form should be of less concern than one discharging mercury which is more easily transformed. However, this assessment is more difficult than simply measuring who is discharging inorganic versus methylmercury. There are a number of species of mercury, other than methylmercury, that may be considered "bioavailable." Some of these may be bioaccumulated in their own right and some may be precursors to methylmercury formation. The presence or absence of these species are determined by the biogeochemistry of effluent waters and receiving waterbody types (i.e. fresh, brackish, marine), individual waterbody characteristics (i.e. pH, dissolved organic carbon [DOC] levels, dissolved oxygen [DO] levels, etc.) and the aquatic foodwebs present to move bioaccumulative mercury up the food chain.

This project was divided into two phases, and this document reports the results of Phase 1. The objectives of this first phase of the project were to:

- develop a working definition of bioavailable mercury;
- identify factors that enhance or mitigate production of bioavailable mercury in aquatic environments;
- profile and rank sources of mercury with respect to bioavailable mercury, enhancers, and mitigators;
- apply the ranking procedure to a variety of sources in a "proof-of-concept" fashion;
- assess the state of the science with respect to those processes that are being used, and may potentially be used, by wastewater treatment plants (WWTPs) to treat and remove mercury from their discharges, and

- Assess the potential regulatory concerns of WWTP owners and operators vis-à-vis the mercury TMDL process.

Bioavailable Mercury Defined

The assessment of the bioavailability of mercury (Hg), and its most toxic and bioavailable organic form, methylmercury (MeHg) is complex. This is especially true in media such as wastewater treatment plant (WWTP) effluents and their receiving waters, as there are numerous factors that can affect the speciation and complexation of Hg and MeHg, and therefore their bioavailability. Because MeHg is biomagnified through all levels of the food chain, it is of primary concern in terms of bioaccumulation into organisms. In addition, however, ionic complexes of inorganic Hg^{II} can be readily converted into MeHg in aquatic systems, as they are bioavailable to the methylating organisms. Thus, in the context of this study, two aspects of bioavailability need to be considered; bioavailability for direct uptake by higher trophic level aquatic organisms, which is primarily related to the MeHg concentration, and bioavailability of inorganic Hg to methylating organisms.

Finally, as the speciation of Hg and MeHg in effluents can change upon mixing or through changes in the overall chemistry of these waters, it is also necessary to consider that some of the Hg and MeHg within the effluents is *potentially* bioavailable given the right conditions in the receiving waters. Given these factors, definitions of bioavailability have been adopted.

Bioavailable mercury *includes those forms of mercury, both inorganic and organic, that can be readily transported across biological membranes, either by active or passive processes and subsequently accumulated, and/or biotransformed to a more toxic and bioaccumulative species.*

Potentially bioavailable mercury *is any form of mercury that can be converted into bioavailable mercury in a scientifically reasonable time span.*

The "scientifically reasonable time span" is postulated to be that period of time until the Hg and MeHg in WWTP effluents (or other sources) becomes indistinguishable in form, speciation, and reactivity from the Hg and MeHg in the receiving waters. The time span is defined to account for any rapid re-equilibration of complexes after mixing of sources and receiving waters, but to exclude any long term changes in speciation. Available evidence on the rate of dissociation of inorganic Hg from complexes with dissolved macromolecular organic matter and from the "strong, stable" complexes found in WWTP effluents and other waters suggests that long term speciation changes will not occur on the timescale of days. Thus, in terms of the bioavailability of Hg in WWTP effluent and, upon mixing, in the receiving waters, the consensus view is that a reasonable time span is of the order of days rather than weeks, but that it is not a constant value for all WWTP effluents or receiving waters, as it is primarily dependent of the nature and strength of the binding of Hg and MeHg to dissolved, particulate and/or colloidal material in the effluent stream. While the data on the variability of Hg binding to WWTP solids is very limited, there is sufficient evidence to conclude that it depends in varying degrees to processes within the plant and on the strength of partitioning of Hg to the solid phase.

The formation of strong Hg-S-NOM associations are considered to be a major determinant of Hg bioavailability. While HgS and other neutral Hg species have been shown to be a source of bioavailable Hg for microbes that are involved in the formation of MeHg, the formation of Hg-S-NOM associations is likely to strongly inhibit bioavailability. In this report, the researchers focused on these reactions as a key factor in the determination of the bioavailability of inorganic Hg.

Certain conventions have been adopted herein to discuss bioavailability and bioavailability concepts. There are three important terms which are essential for the reader to understand and remain aware of when reading this report; specifically, bioavailable fraction, bioavailable concentration, and bioavailable load. The "bioavailable fraction" is the sum of the aqueous bioavailable mercury species relative to total mercury and is dimensionless or expressed as a percent. In general, this term is used interchangeably with "bioavailability" in this report. The use of the bioavailable fraction facilitates important comparisons of various mercury effluents or sources where total concentrations in these sources may substantially differ. The "bioavailable concentration" is the sum of the bioavailable forms of mercury in a water sample expressed in conventional concentration (mass/volume) in units such as ng/L. There is an important distinction between these two forms. For instance, a wastewater treatment process may increase the bioavailable fraction but may reduce the overall bioavailable concentration of mercury by decreasing the total Hg concentration. Thus a process may increase bioavailability but at the same time dramatically decrease the concentration of bioavailable mercury (and therefore the effluent's potential impact on a receiving waterbody). Finally, it is important when comparing the impact of different sources on a receiving waterbody that the bioavailable load be considered. The "bioavailable load" is the product of the bioaavailable concentration with the flow or flux rate from the source into the receiving water. Obviously, a source with a larger bioavailable load will impact a waterbody to a greater extent. For instance, while the bioavailable mercury in sediment porewater may be high, the bioavailable load from the sediments to the receiving water may be low due to limited advective or diffusive flux from the sediments to the water column.

Bioavailability Enhancers and Mitigators

The interaction of factors that may increase or decrease the bioavailability of Hg and MeHg in effluents and receiving waters is also complex. Much of the information needed for a thorough examination of the topic has only been recently collected or has been obtained with specific targeted research studies rather than from long-term monitoring data. Even so, there is a rapidly growing body of knowledge concerned with the factors that affect mercury methylation and bioaccumulation in natural waters.

Table ES-1 below provides a summary of what the researchers believe to be the most important factors enhancing or mitigating the bioavailability of total Hg and MeHg from mercury sources. The relative importance of each parameter is indicated as high, medium, low, no effect (H, M, L, N). In those cases where the issue is too complex for a particular impact to be estimated with certainty, the impact is indicated as uncertain (U). In each case, when considering the impact of a particular parameter, it is assumed that changes in this parameter occur without any concomitant changes in the other factors (i.e. all else remains equal). The details of how these conclusions were reached are dealt with in depth in the report.

Bioavailable and Potentially Bioavailable Mercury Source Profiles

When thinking of bioavailable Hg, it is normal to think of MeHg. However, as pointed out above, it was necessary to expand this definition to include bioavailable and potentially bioavailable inorganic Hg. The need to consider inorganic Hg arises from the fact that in many ecosystems that have been studied, in situ methylation and production of MeHg is the major source in the receiving water rather than input of MeHg from sources. In terms of the sources profiled in the report, it was possible in general to develop characteristics of each with regard to MeHg. However, for bioavailable or potentially bioavailable inorganic species, more often than not this was not possible. In a very few instances, "reactive" Hg was determined in studies of a particular source, but this was the extent. Reactive inorganic Hg (R-Hg) as defined in this report includes species of Hg^{II} that are reducible to Hg^0 by the addition of $SnCl_2$, a mild reductant, prior to analysis. These species are mostly in the dissolved phase but can also include R-Hg associated with particles, and include inorganic and labile organic complexes of inorganic Hg. As noted in the report, this technique does not measure the MeHg.

Table ES-2 attempts to compare %R-Hg and %MeHg for the various sources reviewed and summarized in the report. While it is difficult to place a great deal of certainty in these comparisons, the researchers believe that some logical characterizations have emerged for MeHg. For R-Hg, there is little information and the researchers' characterizations are much less certain. Again, however, certain trends and relationships appear evident. With regard to several sources, there is enough evidence to suggest that R-Hg levels are substantially greater than MeHg levels.

It appears that mining sources have among the lowest %MeHg levels, typically 1% or less. This may be due to a variety of environmental conditions that prevail at most of these sites (and their mine drainage waters and sediments), where either high gradient streams produce well-oxygenated waters, or high sulfide levels inhibit methylation, or due to the fact that a high proportion of the Hg in the solid phase is HgS solids (mercury mines), which are relatively insoluble. Additionally, some fraction of the Hg in sediments is likely to be elemental Hg (Hg^0) (gold mines), which would need to be oxidized before it could be methylated. High concentrations of inorganic Hg may inhibit sulfate reducing bacteria or may induce the activity of MeHg degrading organisms.

Table ES-1. Identification of Enhancers/Mitigators of Bioavailability in Effluents and Receiving Waters and the Magnitude of their Potential Effects.

Enhancer/ Mitigator	Predicted influence of enhancer/mitigator on bioavailability of Hg				
	Mechanisms increasing availability	**Importance***		**Mechanisms decreasing bioavailability**	**Importance***
Effluent and Plant					
T-Hg Concentration	Higher T-Hg would lead to increased bioavailable inorganic Hg	H			
MeHg Concentration	Higher MeHg is postulated to reflect higher bioavailability	H			
TSS Concentration				Higher TSS would likely reduce bioavailability due to sorption.	H
Strength of partitioning of T-Hg and MeHg to solids				Stronger binding of T-Hg and MeHg to solids reduces bioavailability to microbes (decreases dissolved concentrations) and to higher trophic organisms (by decreasing the extent of solubilization during solid digestions). Also, decreases T-Hg bioavailability for conversion to MeHg, and MeHg availability for demethylation.	M
Organic content of effluent	Higher organic content of effluent may stimulate sulfate reduction by providing an available carbon source.	M		Higher organic content in solution would decrease bioavailability due to strong complexation.	H
Presence and strength of Hg-ligand complexes				Presence of these complexes (e.g. Hg-S-NOM) would decrease bioavailability.	H
Presence of major inorganic ions (e.g. sulfate, sulfide, chloride)	No evidence suggesting impact, sulfate is not limiting, Cl appears too low to have a complexing effect. However, the presence of sulfide creates opportunities for formation of bioavailable HgS^0 forms.	N (SO_4^{2-}, Cl^-) M (S^{2-})			
Receiving Waters					
Dilution ratio	May increase or decrease depending on chemistry of receiving waters.	U			

Table ES-1. Identification of Enhancers/Mitigators of Bioavailability in Effluents and Receiving Waters and the Magnitude of their Potential Effects (concluded).

Enhancer/ Mitigator	Predicted influence of enhancer/mitigator on bioavailability of Hg				
	Mechanisms increasing availability	**Importance***		**Mechanisms decreasing bioavailability**	**Importance***
Receiving Waters					
Ionic Strength (Conductivity)				Higher ionic strength tends to coagulate particles resulting in more rapid sedimentation, thereby decreasing bioavailability in the water column. Methylation and release from sediments could mitigate this effect.	L
Relative concentration of DOM, TSS and POC				If relatively high, these could diminish the bioavailability of Hg or MeHg added via effluent. Unlikely to be lower than in effluent. High TSS and POC may not decrease bioavailability of MeHg to higher trophic level organisms.	L
Low DO / high BOD	Low oxygen environments and high biological oxygen demand could lead to enhancement of methylation in the receiving waters.	H			
Sulfate concentration	Increasing sulfate in receiving waters may increase methylation potential.	M			
pH	For low pH receiving waters, methylation could be enhanced.	M			
Receiving water geometry	Shallow, slow flowing systems will enhance the transfer of Hg to sediments.	L - M			
Trophic Status				Increased growth rate and biomass leads to lower Hg concentrations in biota (biodilution).	M
Ecological Sensitivity	Some natural waters seem to be susceptible to enhanced bioaccumulation of methylmercury. Lakes>streams>estuaries Oligotrophic>dystrophic> eutrophic	H			

Table ES-2. Characterization of Reactive Inorganic Mercury (R-Hg) and Methylmercury (MeHg) as a Percent of Total Mercury from Mercury Sources.

Source	%R-Hg	%MeHg
Atmospheric Deposition	15 to 95	1 to 5
Mining Sediments	< 5	< 1
Mining Runoff	< 5	< 1
Mercury-containing Sediments	< 2	< 1 to 5
Sediment Pore Waters		
Impacted	Not known	>10%
Unimpacted	<20%	>10%
Runoff - urban	5 to 10	1 to 2
Runoff - nonurban	~15	5 to 15
Wastewater Treatment Plants	5 to 25	1 to 10

Urban runoff appears to have the next lowest levels of %MeHg, although there a relatively few urban runoff studies that have measured MeHg, and confidence in this range is relatively low. Mercury-containing sediments appear to have the next highest levels of %MeHg, typically ranging from < 1 to 5%. Precipitation surprisingly has characteristics levels of %MeHg of 1 up to 5%.

Wastewater treatment plants appear to have the next lowest levels of %MeHg, although they appear to be more variable in this regard than most other sources. As discussed in Chapter 3.0, levels of MeHg in wastewater effluent are undoubtedly impacted greatly by the design and operation of individual plants. In particular, the presence of anaerobic processes in WWTPs (e.g. recycling of sludge digestion supernatant) *may* explain elevated levels in *some* cases, as was suggested in studies reviewed in this report. WWTPs have the potential to have a higher relative concentration of R-Hg due to the typically low solids content in the effluent. However, there is evidence, presented in Chapter 2.0, that WWTP effluents may have a higher percentage of strong Hg binding ligands, which may be a combination of organic matter-sulfide complexes and which may reduce the bioavailability of Hg^{II} in effluents. These complexes may also be quite resistant to degradation, even in oxic receiving environments. Work completed in Phase II of this project shows that for the treatment plants studied, most %R-Hg values fell within a range of 5-25% of total mercury in the effluent.

Non-urban runoff appears to have the highest %MeHg levels, falling for the most part in a range from 5-15%. This latter result may reflect the presence of bogs and wetlands in many natural watersheds which tend to elevate MeHg in this relative to other sources.

Bioavailability Estimation Methods

Definitions of bioavailable mercury are used to construct a framework and general approach to assessing the relative bioavailability of mercury (Hg) in sewage treatment plant effluents and other source waters. The focus is on the bioavailability of Hg in the source waters and this Phase I report does not deal in any detail with the impact of mixing with receiving waters on the bioavailability, which in part will be the business of Phase II. The approach taken is also to define and estimate the fraction that is bioavailable rather than the total amount of bioavailable Hg, as this makes comparison between systems easier and is not reliant of factors such as large differences or variability in total Hg concentrations or loading rates.

The approach taken is based on the assumptions that all the methylmercury (MeHg) in the water is bioavailable and that only a fraction of the dissolved inorganic Hg (that fraction corresponding to R-Hg) is bioavailable, while particulate inorganic Hg is not. Furthermore, it is concluded that the dissolved inorganic Hg bound to dissolved natural organic matter may be considered essentially non-bioavailable. The rationale for these conclusions is detailed in the report. Therefore, the fraction of the total Hg that is bioavailable, denoted as fB-Hg, is equivalent to the fraction as MeHg plus the dissolved reactive Hg (R-Hg) fraction.

If the %MeHg and %R-Hg is known or can be estimated from the literature, then fB-Hg, the fraction of bioavailable mercury, can be estimated directly, as a fraction or as a percentage. While MeHg concentration is not always measured for effluents and other sources waters, there are values in the literature that can allow a reasonable estimation of the fraction of Hg as MeHg. Such an approach was used, for example, in constructing Table ES-3. Based on the available data and the definition of bioavailability as adopted in this report, Table ES-3 shows that mining sources and urban runoff tend to have the lowest fractions of bioavailable mercury. Mercury from wastewater treatment plants also appears to have lower bioavailability compared to other sources evaluated in this report. Sediment pore water and atmospheric deposition appear to be among the highest in terms of mercury bioavailability.

The chapter provides alternative ways of determining each of the method parameters or surrogates for these parameters, and discusses the types of estimations that can be made, and their certainty, which is a function of the amount and extent of the information available. There are other possible approaches to estimating the fB-Hg; however, given the focus of the report, it is not relevant to detail the specifics of other approaches. The methodology outlined in the report is applied to several specific datasets to provide a "proof-of-concept" of this particular approach.

Table ES-3. Estimated Fraction of Dissolved Mercury Considered Bioavailable, Using Ranges in the Estimated Fraction of the Dissolved Mercury that is Reactive, and Ranges in Methylmercury Concentration for Different Source Water Types.

Water Type/Source	%MeHg	%R-Hg	Range in estimated bioavailable Hg fraction
Atmospheric Deposition	1-5	15-95	0.2 - > 0.9
Mining Runoff	0.1-1	1-5	< 0.1
Impacted Sediment Pore water	10-25	5-20	0.15 – 0.45
Unimpacted Sediment Pore water	10-25	10-20	0.2 – 0.45
Runoff - urban	1-2	5-10	< 0.1
Runoff - nonurban	5-15	15	0.2 – 0.3
Wastewater Treatment Plant Effluent	1-10	5-25	< 0.3

Mercury in Wastewater Treatment Plants

The researchers drew the following conclusions with regard to wastewater treatment on mercury in municipal effluents:

- Mercury is very effectively removed by wastewater treatment plants. Depending on the treatment technologies employed, and the influent concentration, from 80% to greater than 99% of the total mercury entering a wastewater plants is removed. Because of the affinity of mercury for particulates, most of the mercury becomes incorporated in biosolids.[1]
- In general, most wastewater treatment plants appear to generate only small quantities of MeHg as part of the treatment process and overall reduce MeHg in effluent relative to influent. However, some plants do appear to have effluent MeHg levels that are elevated with respect to background. The reason for this is unknown, as higher MeHg levels do not seem not to correcpond to type of plant, location, or other identifiable causative factors. Potential sources of MeHg in these plants and control opportunities should be individually investigated.
- Many primary and secondary wastewater treatment processes would appear to enhance the bioavailable mercury fraction in effluent even though, overall, wastewater treatment is very efficient at removing total mercury (as well as bioavailable mercury) from the effluent stream[2]. Enhanced bioavailability is hypothesized to result from efficient removel of solids, thereby promoting a higher percentage of dissolved Hg in effluents relative to influents. However, this potential effect may be strongly mitigated by the presence of both organic and inorganic ligands which bind dissolved Hg and may greatly reduce its bioavailability[3]. Advanced treatment processes that target complexed Hg in solution are presumed to decrease Hg bioavailability.
- The effects of wastewater treatment technologies on the bioavailability of mercury are not well-established and data are lacking in this regard. More information needs to be generated in the following areas:
    - the distribution of mercury in effluents in either the dissolved or adsorbed phase (fraction of dissolved versus particulate mercury);

---

[1] Mercury in biosolids is regulated for land application and incineration by the *Standards for Sewage Sludge Use and Disposal* (40 CFR Part 503). Mercury emissions from incinerators is the subject of current WERF research (04-CTS-7). Recent biosolids surveys indicate that mercury in biosolids is well below the Part 503 limits for land application.

[2] The reader is reminded of definitions adopted by the authors in this context. Usage of the term "bioavailability" connotes the concentrations of bioavailable species relative to the total mercury concentration in a source or effluent. The term "bioavailable mercury" on the other hand connotes the actual concentrations of bioavailable mercury species in a sample. Thus, an effluent may have greater bioavailability (i.e. a higher bioavailable fraction) compared to another source, but have less bioavailable mercury (i.e. lower bioavailable mercury concentrations).

[3] As is pointed out in Chapter 3.0, municipal wastewaters appear to have metal sulfide: DOC complexes that strongly bind Hg and are not removed by filtration. However, there is debate as to the composition of these complexes and there is at present no reliable information as to their effect on the bioavailability of bound mercury. To the extent that they consist of larger DOC: sulfide complexes, they may reduce bioavailability. To the extent that they tend to be simple soluble, neutral inorganic HgS, they may be readily bioavailable. Obviously, there is much needed research on this topic.

- the degree of strong/weak complexation of dissolved mercury by organic and inorganic ligands;
- the fraction of MeHg in effluents; and
- where, and the degree to which, recalcitrant ligands form complexes with mercury and opportunities to alter such complexation.

- Some promising technologies are emerging for advanced treatment of mercury following secondary treatment; however, there continues to be a dearth of information regarding the testing of these technologies at low levels and with municipal wastewaters. *At this time, there are no proven, cost-effective methods to consistently bring typical total Hg levels in municipal wastewater effluent (10 to 100 ng/L) to typical regulatory levels (1 to 5 ng/L) although some plants appear to achieve thse levels.* The most promising of these emerging technologies are either chemical or biological processes (i.e. addition of chemical reductants, photoreduction, and microbial reduction) that involve reduction of Hg^{2+} to Hg^0 and the subsequent volatilization or stripping of Hg^0 from the aqueous media. However, the loss of this Hg to the atmosphere may not be desirable from a regulatory standpoint. The costs of these technologies are not well-established; however, the cost of biological treatment appears to be relatively high. Furthermore, most of these technologies have been evaluated on industrial wastewaters having very different characteristics from municipal effluents, and at much higher concentrations. As such, any technologies of interest should be tested on municipal effluent to properly evaluate treatment potential and cost.

TMDL Allocations

The report summarizes key findings from the TMDL reviews. It is based upon review of the 73 TMDLs as described in EPRI (2006) and the more recent TMDLs reviewed for this project. Since the review of more recent TMDLs has added little in the way of new information, these findings are taken largely from the EPRI report. Findings are summarized in four categories:

- Use of Fish Tissue Criterion as a Water Quality Standard
- Water Column Targets Derived from Fish Tissue Targets
- Approaches for Deriving Water Column-Based Criteria
- Approaches to TMDL Allocations and Implementation

Mercury TMDLs have been and will continue to exert an influence on wastewater treatment in the United States. This becomes obvious when comparing the typical WWTP effluent with the water quality targets established through the TMDL process. Comparison of effluent discharge levels (which range up to several hundred ng/L) with typical T-Hg water quality targets established by mercury TMDLs across the country (0.5 to 41 ng/L) indicates that in a given waterbody, an individual WWTP may have difficulty being in compliance if water quality targets for T-Hg are adopted as end-of-pipe permit levels. The same is true for MeHg, where effluent concentrations may range as high as 2 to 3 ng/L and typical TMDL water quality targets are typically less than 0.5 ng/L. A complicating factor is that treatment of wastewater to "single-digit" ng/L levels or lower is expensive and in some cases may not technologically feasible. It is still not clear why some plants appear to achieve mercury removal to these levels with conventional technologies while other plants with similar treatment processes do not. Most emerging technologies for removal of mercury have not been tested in the high-volume low concentration environments of municipal WWTPs and are unproven in this context.

CHAPTER 1.0

INTRODUCTION

Mercury levels in fish tissue are considered by regulatory agencies to be a widespread human health concern, and fish consumption advisories have been issued for over 2000 waterbodies in the United States. The existence of fish consumption advisories in a waterbody have in turn been used as a sufficient basis for the listing of these waterbodies on the states' lists of impaired waters, the 303(d) lists. The sources of mercury are known to be diverse, as it is a naturally occurring element, although it is generally accepted by the scientific community that the emissions of mercury have increased several fold during the industrial period. Therefore, attempts to regulate mercury discharges to waterbodies have focused on point source dischargers and air emissions.

The TMDL process has been used to attempt to quantify mercury loadings to waterbodies – both to estimate the acceptable load that would result in the attainment of acceptable levels in fish tissue, and to estimate the current load of mercury to the waterbody that is resulting in nonattainment. In the case of mercury, a particularly difficult aspect of this exercise arises from the fact that it must be transformed from the form that is typically deposited or discharged to waterbodies (inorganic, primarily divalent mercury) to the form that is bioaccumulated in fish (methylmercury). Currently, there are no accepted techniques available to "trace" mercury from its emitted origin to its appearance in fish tissue[4]. Therefore, the question of "whose mercury is it?" remains unanswered.

The movement toward an assessment of the bioavailability of various mercury sources is a step in this direction and is a rational one. It stands to reason that a source which may be contributing mercury that is more difficult to transform into a bioaccumulative form should receive less attention than one discharging mercury that is more easily transformed. However, this assessment is more difficult than simply measuring who is discharging inorganic versus methylmercury. There are a number of species of mercury, other than methylmercury, that may be considered "bioavailable". Some of these may be bioaccumulated in their own right and some may be precursors to methylmercury formation. The presence or absence of these species are determined by the biogeochemistry of effluent waters and receiving waterbody types (i.e. fresh, brackish, marine), individual waterbody characteristics (i.e. pH, dissolved organic carbon [DOC] levels, dissolved oxygen [DO] levels, etc.) and the aquatic foodwebs present to move bioaccumulative mercury up the food chain.

1.1 Purpose of Research

1.1.1 Background

The motivation for this study grew out of the San Francisco Bay mercury Total Maximum Daily Load (TMDL). As part of that process it was estimated that the wastewater treatment plants (WWTPs) discharging into the Bay accounted for only a small percentage of the estimated total mercury entering the Bay on an annual basis (about 2.5%). However, questions were raised concerning the bioavailability of mercury from the WWTPs versus other sources of mercury including stormwater, mining sources,

[4] Although stable isotopes of mercury have been used in scientific studies such as METTALICUS (Harris et al., 2007) to attempt to trace the movement of mercury in watersheds and aquatic foodwebs under experimental conditions, there are still no distinctive isotopic "signatures" for the various mercury sources that would allow such a technique to be of widespread utility in a regulatory context.

contaminated sediments and air deposition, to name a few. The WWTPs were thus obligated to perform a study to make an assessment of this type. WERF became aware of this research challenge and made the decision to host the project. Additional funding was provided by the Bay Area Clean Water Agency (BACWA).

This was a prudent course of action for several reasons:

- WERF sponsorship lent a high level of objectivity and legitimacy to the study and study results.
- WERF sponsorship elevated the study to a nationwide effort, as many other WWTPs in various parts of the country will inevitably face similar challenges.
- The project geography and the types of receiving water environments were expanded so that the study would be useful to a wider population of the WERF membership.

1.1.2 Objectives

1.1.2.1 Phase 1

This project was divided into two phases and this document reports the results of Phase 1. The objectives of this first phase of the project were to:

- develop a working definition of bioavailable mercury (Chapter 2.0);
- identify factors that enhance or mitigate production of bioavailable mercury in aquatic environments (Chapter 3.0);
- profile and rank sources of mercury with respect to bioavailable Hg, enhancers, and mitigators (Chapters 4.0 and 5.0);
- apply the ranking procedure in a "proof-of-concept" fashion (Chapter 5);
- assess the state of the science with respect to those processes that are being used, and may potentially be used, by WWTPs to treat and remove mercury from their discharges (Chapter 6.0); and
- assess the potential risks to WWTP owners and operators vis-à-vis the mercury TMDL process (Chapter 7.0).

The development of a working definition of bioavailable mercury was critical to the success of this project. It was important that the definition of bioavailable mercury not only included methylmercury, the form bioaccumulated by higher trophic-level organisms, but also the forms that are bioavailable to organisms that methylate mercury. The definition in turn must not only encompass the mercury species that are immediately bioavailable but also those *potentially* bioavailable within a reasonable time-scale. Only at this level is the definition broad enough to account for biogeochemical transformations that occur from the discharge point to the end point of methylmercury in fish tissue in various receiving water environments.

It is also essential to identify the major environmental factors that enable the transformations of the species in the pool of labile bioavailable mercury to bioavailable mercury species (enhancers) and those that mitigate or retard these processes (mitigators). These factors will be different in different receiving water environments and in the various mercury sources to a given waterbody. To this end it is also essential that the major mercury sources are identified and their effluents profiled with regard to bioavailable and potentially bioavailable species discharged, along with those substances that potentially enhance or mitigate transformations from the labile to the bioaccumulative pools. In Phase 1, a ranking protocol was developed to enable a "relative bioavailability" scoring for effluents from selected sources.

Once the methodology for ranking various effluents was developed, it was important to use it in one or more "real-world" situations in order to determine its usability, its relevance, and to assess whether the data required to use it are generally available to potential users.

A literature review was performed to assess the effectiveness of currently deployed technology to treat mercury in wastewater effluents. In addition, emerging technologies that might be applicable for mercury treatment in WWTPs were reviewed, both in terms of effectiveness and cost. Finally, to make a connection to the assessment of bioavailability, these various processes and technologies were reviewed in terms of their potential effect on mercury bioavailability. Finally, an assessment of mercury TMDLs was provided so the potential risks to wastewater treatment plant owners and operators could be articulated.

1.1.2.2 Phase 2

It is believed that readers of this report will best be served by the development of a two-tiered guidance document in Phase 2. One tier will be a screening level assessment tool that builds upon the ranking protocol developed in Phase 1. This tool will be most useful in assessing overall risk relative to other dischargers in a watershed, making the decision to perform a more detailed assessment, and allocation of resources if a more detailed assessment is warranted. The second tier will be a set of guidelines and protocols to conduct a detailed-level assessment, including sampling, analysis, data analysis and interpretation, to make a more definitive assessment of bioavailability in WWTP effluents. Such an assessment may be warranted if the WWTP is directly involved in a TMDL or similar regulatory action and is asked to make a demonstration of the bioavailability (or lack thereof) of mercury in its effluent and mixtures of its effluent with the receiving water, taking into account both the characteristics of the effluent and the characteristics of the receiving waterbody.

CHAPTER 2.0

MERCURY BIOAVAILABILITY

2.1 Summary

The assessment of the bioavailability of mercury (Hg), and its most toxic and bioavailable organic form, methylmercury (MeHg), is complex. This is especially true in media such as wastewater treatment plant (WWTP) effluents and their receiving waters, as there are numerous factors that can affect the speciation and complexation of Hg and MeHg, and therefore their bioavailability. Because MeHg is biomagnified through all levels of the food chain, it is of primary concern in terms of bioaccumulation into organisms. In addition, however, ionic complexes of inorganic divalent Hg (Hg^{II}) can be readily converted into MeHg in aquatic systems, as they are bioavailable to the methylating organisms. Thus, these complexes of inorganic mercury also need to be considered as bioavailable for the purposes of this project.

Thus, in the context of this study, two aspects of bioavailability need to be considered:

- Bioavailability for direct uptake by higher trophic level aquatic organisms, which is primarily related to the MeHg concentration
- Bioavailability of inorganic Hg to methylating organisms

Finally, as the speciation of Hg and MeHg in effluents can change upon mixing or through changes in the overall chemistry of these waters, it is also necessary to consider that some of the Hg and MeHg within the effluents is *potentially* bioavailable given the right conditions in the receiving waters. Given these factors, definitions of bioavailability were adopted.

Bioavailable mercury *includes those forms of mercury, both inorganic and organic, that can be readily transported across biological membranes, either by active or passive processes and subsequently accumulated, and/or biotransformed to a more toxic and bioaccumulative species.*

Potentially bioavailable mercury *is any form of mercury that can be converted into bioavailable mercury in a scientifically reasonable time span.*

The "scientifically reasonable time span" is postulated to be that period of time until the Hg and MeHg in WWTP effluents (or other sources) becomes indistinguishable in form, speciation, and reactivity from the Hg and MeHg in the receiving waters. In terms of the bioavailability of Hg in WWTP effluent and, upon mixing, in the receiving waters, the consensus view is that the time span is of the order of days rather than weeks, but that it is not a constant value for all WWTP effluents or receiving waters, as it is primarily dependent of the nature and strength of the binding of Hg and MeHg to dissolved, particulate and/or colloidal material in the effluent stream. While the data on the variability of Hg binding to WWTP solids is very limited, there is sufficient evidence to conclude that it depends in varying degrees to processes within the plant and on the strength of partitioning of Hg to the solid phase.

It is reasonable to consider only ionic Hg in the +2 oxidation state (Hg^{II}) and MeHg when assessing potential bioavailability. Furthermore, only the dissolved forms of Hg^{II} and MeHg can be bioaccumulated by organisms at the base of the food chain that can transform Hg^{II} into MeHg. The major factor controlling the inorganic speciation is chloride (Cl^-) complexation as both Hg and MeHg form strong complexes with Cl^-. In the absence of high Cl content, and also at high pH, the hydroxide complexes are dominant. In addition to the inorganic complexes such as chloride and hydroxide (e.g. $HgCl_2$, $Hg(OH)_2$), the ionic Hg^{II} pool includes complexes of organic ligands, such as cysteine, EDTA and NOM.

The most important components of effluent and receiving water chemistry that determine the bioavailability to the base of the food chain are the amount and type or organic matter (NOM) and the presence of filterable sulfide-containing compounds. Both of these range in size from truly dissolved molecules to colloidal material that pass through conventional filtration systems. In most situations, the presence, type and concentration of NOM is the primary determinant of the Hg and MeHg speciation in the filtered fraction. For WWTP effluents, it has also been shown that there are strong metal binding sulfide-containing compounds that can effectively bind the Hg to make it relatively unreactive and that such compounds can persist in the environment even under conditions where they may not be thermodynamically stable. The presence of such compounds will have a dramatic effect on the bioavailability of Hg and MeHg, and also on the potential for Hg^{II} to be converted into MeHg.

There is an increasing body of research demonstrating the ability of the neutrally-charged complexes of both Hg and MeHg to cross biological membranes by passive diffusion. This has been demonstrated for the uptake of the neutral chloride complexes – $HgCl_2^0$ and CH_3HgCl^0 – by phytoplankton, and for the neutral complexes of Hg with sulfide – HgS^0 and $Hg(SH)_2^0$ – by methylating bacteria.

However, it should be noted that it is possible for both Hg^{II} and MeHg to be assimilated from the solid phase by higher trophic level organisms such as benthic invertebrates and filter feeders which feed on sediment and/or suspended solids (TSS) which includes particulate organic matter (POM). For these organisms, the strength of binding of the Hg^{II} and MeHg to the solid phase determines bioavailability, and it has been shown that bioavailability and bioaccumulation is inversely related to the sediment concentration of natural organic matter (NOM). Thus, Hg^{II} and MeHg in sediments with a high NOM are much less bioavailable than sediments with low NOM. While it is true that sediments with a high NOM typically contain more Hg and MeHg than low NOM sediments, it can be concluded that overall the Hg^{II} and MeHg will be accumulated into benthic organisms to a lesser degree from high NOM sediments.

Overall, the bioavailability of Hg and MeHg in WWTP effluents and receiving waters can be assessed if a number of parameters are known or can be assayed. The primary factors are the organic content of the waters and solids, the total concentrations of Hg and MeHg and the relative MeHg concentration (%MeHg), the TSS concentration, the presence of "filterable" sulfide-containing compounds, and the strength of binding of Hg and MeHg to the solids. Other factors are important and these will be discussed further in the next chapter, which deals with the factors that enhance or mitigate Hg bioavailability.

2.2 Introduction

While the issue of defining mercury (Hg) and methylmercury (MeHg) bioavailability is complex, important strides have been made in the last few years. Therefore it is possible to more accurately define the bioavailability of Hg in all its forms in different media, and, for the purpose of this study, the bioavailability in wastewater treatment plant (WWTP) and other effluents. It is necessary first to examine the terminology and define the relative bioavailability of Hg in all its forms.

Inorganic Hg exists in two primary oxidation states (two and zero) in most aquatic waters – as Hg^{II} complexes in solution or attached to particles, and as a dissolved gas, elemental Hg (Hg^0). Note that the superscripts 0, I, II (positive signs will be not be included) will be used to represent the various oxidation states of Hg and MeHg while the superscripts 0, +1 (or simply +), +2 will be used to represent the actual charge on the molecule or complex. For example, Hg^{2+} refers specifically to the free metal ion while Hg^{II} refers to all ionic forms of Hg in the +II oxidation state; similarly, CH_3Hg^+ refers to the free ion of MeHg. As discussed below, for inorganic Hg, it is only Hg^{II} that is bioaccumulative in aquatic systems, and this is the primary form of inorganic Hg in most aquatic environments of interest to this study (Mason, 2002; Morel, et al., 1998). In most aquatic systems, the free ion concentration (Hg^{2+}) is a very small fraction (<<1%) of the total dissolved Hg^{II}, so it is the inorganic complexes (e.g. $Hg(OH)_2$, $HgCl_3^-$, $HgCl_2$, $Hg(SH)_2$) and organic complexes (mostly Hg^{II} bound to NOM) that determine/control the fate and bioavailability of Hg^{II} to organisms. Furthermore, only dissolved species are thought to be taken up by microorganisms (bacteria, phytoplankton) while higher trophic level organisms primarily obtain their Hg and MeHg from ingestion of food, sediment or other "particles" (e.g. Wiener et al., 2003). Thus, the issue of bioavailability of sediment-bound Hg and MeHg, or that in the suspended solids, needs to be considered for higher trophic level organisms feeding directly on the sediment or ingesting solids.

The term "organic Hg" is mostly used to refer to compounds where an organic group is covalently bonded to Hg, such as with MeHg (i.e. CH_3Hg^+ ion), dimethylmercury ($(CH_3)_2Hg$) or ethylmercury ($CH_3CH_2Hg^+$). Note that the oxidation state of Hg in these forms is still +II, the same as that of ionic Hg, but that the charge for MeHg on the free metal ion is +1 (i.e. CH_3Hg^+ and CH_3Hg^{II} complexes). Also, for the species that are charged, these compounds have the ability to form both inorganic and organic complexes in solution. In the environment, MeHg is the dominant organic Hg form and will be one of the focuses of this study. Thus, there is a distinction made between compounds where there is an organic moiety covalently bonded to the Hg atom and those where the association is "more ionic" and can be thought of as a complex of Hg^{II}, with the organic compound acting as a ligand. This separation and definition is somewhat arbitrary as, for example, Hg forms "relatively covalent" bonds in many of its complexes. These complexes are, however, more labile in that the exchange of ligands occurs relatively rapidly in natural waters. In contrast, the addition and removal of the methyl group from CH_3Hg^+ does not occur readily although such "demethylation" processes occur in environmental waters and in sediments/soils, and are both abiotically (photochemically) and biotically mediated. The reverse process of "methylation" is thought to be primarily biotic in the environment (e.g. Benoit et al., 2003).

U.S. EPA's website lists a number of definitions of bioavailability and selects this definition: *"A measure of the physicochemical access that a toxicant has to the biological processes of an organism. The less the bioavailability of a toxicant, the less its toxic effect on an organism."*

For mercury, however, this definition does not encompass the entire spectrum of bioavailability, as toxicity is not the only, or even the primary, issue. While it is important to consider the bioavailability of inorganic Hg and MeHg within the aquatic system and its availability and impact on higher trophic level organisms, the bioavailability of primarily inorganic (ionic) Hg to processes (mostly biotic transformation to MeHg) that convert it to more toxic and bioaccumulative forms, principally MeHg, must also be considered. For most aquatic organisms, except in highly contaminated environments, toxicity of Hg is not the primary issue and thus the emphasis is on bioaccumulation rather than toxicity. Furthermore, for higher trophic level organisms, MeHg is the primary concern as it is much more bioaccumulative and is biomagnified at each level of the food chain. This is not so for inorganic Hg, which behaves in a similar fashion to other Group 12 metals (cadmium and zinc) in terms of its trophic transfer (Mason, 2002).

These aspects of Hg and MeHg biogeochemistry need to be all considered if one is to define bioavailability in a comprehensive sense. Therefore, it is necessary to provide a working definition and then examine current knowledge in light of this definition and approach, and how various interpretations of the data may alter this definition.

2.3 Working Definition of Bioavailability

For the purposes of this project, the U.S. EPA definition is modified to the following overall working definition of the *bioavailable Hg and MeHg for bioaccumulation and biotransformation.* This term refers to that portion of the Hg that can be easily assimilated by the microorganisms at the base of the food chain, and, as a result, by higher trophic level organisms, and also that Hg that is accumulated and subsequently biotransformed into more bioavailable and bioaccumulative forms. Therefore, the research team's working definition is:

Bioavailable mercury includes those forms of mercury, both inorganic and organic, that can be readily transported across biological membranes, either by active or passive processes and subsequently accumulated, and/or biotransformed to a more toxic and bioaccumulative species.

This definition is that used, for example, by Mason et al. (1996) to assess the bioavailability and toxicity of Hg and MeHg to a coastal diatom, and that used by Benoit et al. (1999a; 1999b) in their methylation studies with bacteria. In this form, the definition encompasses all oxidation states (i.e. of Hg^0, Hg^I and Hg^{II}) and also all inorganic Hg and organic Hg complexes. However, as noted above, there are specific forms and oxidation states that will be the focus of the study. Note further, as discussed below, that a form of Hg not currently bioavailable can over time be transformed into a bioavailable form. Therefore it is necessary to stipulate that the term bioavailable Hg represents bioavailability of Hg and MeHg in its current form. In addition, it is also necessary to provide a working definition of the term *potentially bioavailable Hg and MeHg*:

Potentially bioavailable mercury is any form of mercury that can be converted into bioavailable mercury in a scientifically reasonable time span.

The definition of a reasonable time span is a crucial aspect of the discussions concerning Hg bioavailability, and is one of the primary issues that needs to be addressed and which needs further study. This time span is postulated to be the period of time until the Hg and MeHg in

WWTP effluents becomes indistinguishable in form, speciation, and reactivity from the Hg and MeHg in the receiving waters. Currently this is an issue of debate but the research team proposes, based on the available research and knowledge of Hg and MeHg aquatic chemistry, that this timescale is of the order of days, rather than weeks or months, for dissolved Hg and MeHg in effluents. It has been shown that different complexes of Hg^{II} have different rate coefficients for exchange reactions with Hg bound to complex organic ligands, reacting much slower than when bound to simpler organic compounds or inorganic ligands (Lamborg et al., 2003). The noted high reactivity of the inorganic complexes is consistent with theory (Morel and Hering, 1993). Also, it appears that some forms of Hg in the dissolved fraction of WWTP effluents are relative stable and do not readily decompose under conditions where such a change is thermodynamically predicted (e.g. in the presence of another ligand; Hsu-Kim and Sedlak, 2005).

The possibility that ionic Hg can be converted into CH_3Hg, the most bioavailable and toxic form of Hg and *vice versa* by methylation and demethylation (Figure 2-1; Benoit et al., 2003), also requires a consideration of how and when such transformations influence bioavailability, as indeed they do. Furthermore, there needs to be consideration of the potential for Hg and MeHg released in the solid phase from WWTPs to become bioavailable at the base of the food chain by microorganisms and/or transformation after its release into solution. For higher trophic level organisms, there is the potential for the assimilation of Hg and MeHg from the sediments during digestion which needs to be considered as well. Overall, as discussed below, there is an inverse relationship between the bioavailability of Hg and MeHg in sediments and the sediment NOM so that Hg and MeHg is relatively less bioavailable in high NOM sediments.

2.4 Background and Rationale for the Definition

2.4.1 Inorganic Mercury Speciation and Bioavailability

As noted briefly above, considering the various oxidation states of Hg, it appears that Hg^0 is not readily accumulated into microorganisms in aquatic systems. However, Hg^0 will be accumulated if it is oxidized after transport across the membranes – as happens, for example, in

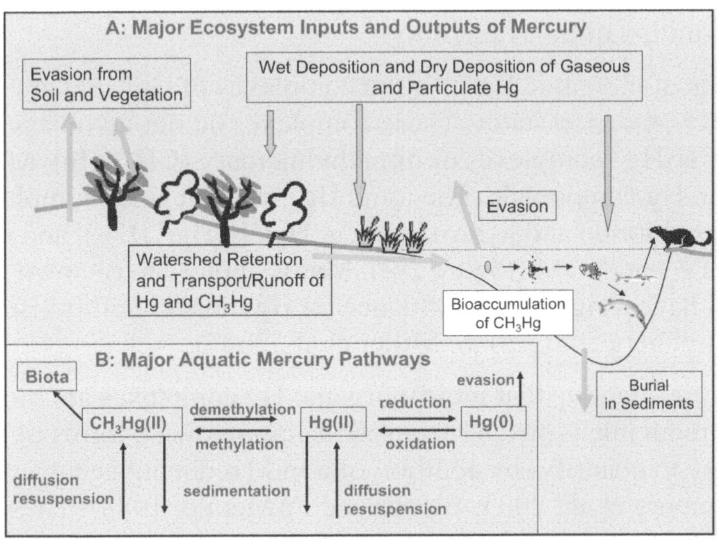

Figure 2-1. Major Environmental Pathways for Mercury in the Environment.
Reprinted with permission from Mason et al., *Environmental Science & Technology*, American Chemical Society (2005).

the blood of humans (PerrinNadif et al., 1996). As discussed further below, such a transformation is not likely within the cell of a microorganism (Mason et al., 1996) and thus Hg^0 is not considered bioavailable according to the definition above. Of the two other oxidation states in which Hg can exist, the +1 oxidation state (Hg^I) is rarely found in aquatic systems except under very unique conditions – not likely to be encountered in the effluent of WWTPs. Thus, this oxidation state is not considered further.

The transformations of methylation and demethylation can be biologically mediated, and such processes have been shown to occur within the cells of microorganisms rather than on the cell surface (Benoit et al., 2003). Therefore, based on the definition above, Hg compounds need to be bioavailable to be transformed by biotic processes. However, it is known that CH_3Hg can also be demethylated photochemically (abiotically) (Sellers et al., 1995) and such a transformation converts CH_3Hg into a bioavailable or non-available form, depending on the end-product - Hg^{II} species in solution or Hg^0. The product of photochemical degradation of CH_3Hg is an avenue of active research, but the general trends appear to be known; that is, that such processes are hindered in colored waters or waters of high suspended solids concentration as both limit UV light penetration. Thus, the potential for this process to be important can be determined based on the effluent and receiving water characteristics.

Additionally, there is the potential for redox transformation in the water column – both oxidation of Hg^0 to Hg^{II} and reduction of Hg^{II} to Hg^0 (Poulain et al., 2004; Garcia et al., 2005; Whalin and Mason, 2006) and clearly these transformations have an important impact on the bioavailability of the Hg. Thus, such transformations need to be considered part of the processes involved in transforming potentially bioavailable Hg into bioavailable Hg^{II}. However, they can result in the reverse situation as well – removal of bioavailable Hg, especially as Hg^0, a dissolved and volatile gas, can be lost from the water column via gas exchange (Figure 2-1). Rates of reduction in surface waters are typically of the same order as rates of evasion. These occur such that a relatively small fraction of the effluent Hg would be lost via reduction and evasion under the most likely conditions encountered. However, in considering all the potential transformations, the issue of the time span over which they occur is crucial and is an integral part of the researchers' refined definition of bioavailability.

To summarize, in terms of inorganic Hg, it is the complexes of Hg^{II} that are the focus of consideration, and not the other oxidation states. These complexes do not involve a covalently bonded organic moiety (e.g. CH_3Hg^{II} complexes or dimethylmercury ($(CH_3)_2Hg$), as such compounds are termed organic Hg compounds. The ionic Hg^{II} pool includes complexes with both inorganic ligands, such as chloride and hydroxide (e.g. $HgCl_2$, $Hg(OH)_2$), and organic ligands, such as cysteine, EDTA and NOM (Figure 2-2). Many studies have shown that Hg binds strongly to NOM and this can have a significant influence on Hg bioavailability (Benoit et al., 2001; Ravichandran, 2004; Skyllberg et al., 2006; Miller et al., 2007).

A review of the literature suggests that most inorganic Hg complexes are bioavailable as they are "reactive" or "easily reducible" - two terms used to classify those forms of Hg that are reducible to Hg^0 (which is easy to quantify) by addition of a mild reducing agent, acidic tin(II) chloride solution ($SnCl_2$) (Lamborg et al., 2003; Bloom and Fitzgerald, 1988). These definitions are based on a chemical reaction and a specific analytical method. As noted elegantly by Lamborg (2006), the definition of *reactive Hg* is that of species in solution, or attached to particulates, that can react readily on the timescale of the measurement (typically 15-30 minutes) and therefore represents those forms of Hg in the sample that are chemically labile, or readily bioavailable. This chemical method provides a reasonable measure of the bioavailable ionic Hg^{II} pool, and this has been proposed in the past (Mason, 1991; Fitzgerald et al., 1991), but it is limited in that it does not assay bioavailable MeHg (i.e. dissolved CH_3Hg^{II} compounds). Additionally, any Hg^0 within the waters are also quantified in this fraction, and needs to be accounted for, most effectively by stripping Hg^0 from solution prior to the reactive Hg analysis. Given these concerns, the usefulness of the "reactive Hg" approach is limited and confined to the examination of ionic Hg only. The mild reducing agent $SnCl_2$ does not convert CH_3Hg^{II} compounds to Hg^0 and is therefore not a suitable measure of the bioavailable CH_3Hg in a sample (Mason, 1991; Bloom and Fitzgerald, 1988).

In terms of complexation in oxic waters, inorganic Hg and MeHg speciation is different in saline versus fresh water environments. The major factor controlling the speciation is chloride (Cl^-) complexation as both Hg and MeHg form strong complexes with Cl^- (Figure 2-3). In the absence of high Cl content, and also at high pH, the hydroxide complexes are dominant. Other inorganic complexes that may be important in some instances are complexes with carbonate and phosphate. Overall, however, in the environment, the two major factors influencing the extent and importance of inorganic complexes are the concentration of Cl and the pH of the system. This is shown, for example, in Figure 2-3, for changing Cl concentration ($pCl = -\log[Cl^-]$) at a pH of 6. As noted above, CH_3Hg^+ is a minor component at low Cl concentration (high pCl) while Hg^{2+} is not important at any pCl or pH (Mason et al., 1996). Thus, in the absence of organic ligands, the hydroxide complexes dominate in freshwater and chloride complexes in seawater.

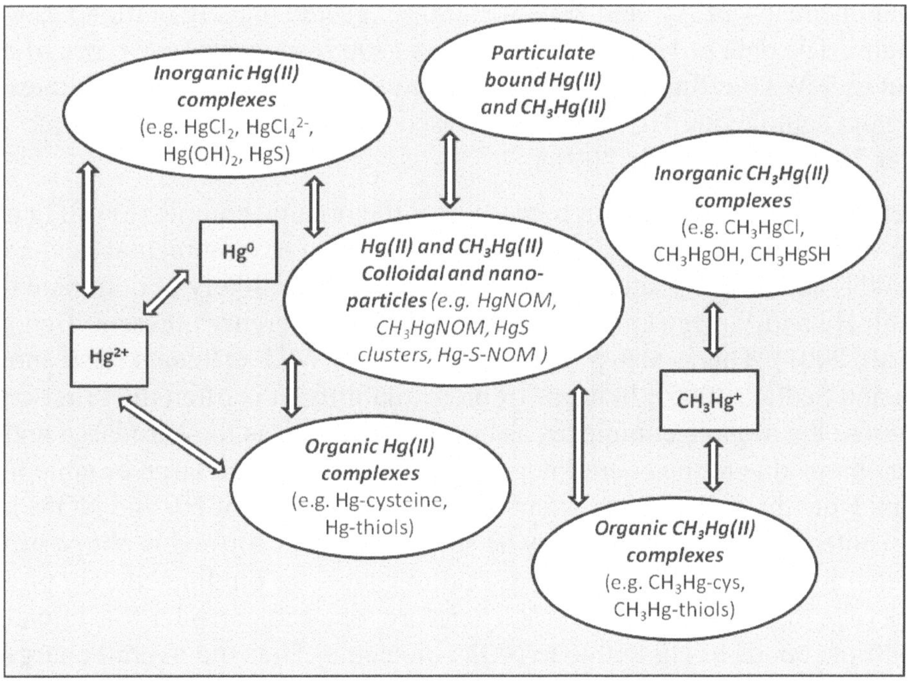

Figure 2-2. Schematic Presentation of the Potential Interactions between the Various Forms of Mercury and Methylmercury in Natural Waters.

The factors influencing complexation of Hg^{II} by inorganic ligands are well quantified and understood, and the same is true for Hg and MeHg binding to NOM (e.g. Ravichandran 2004; Miller et al., 2007; Hsu-Kim and Sedlak, 2005). Additionally, there is the potential for Hg^{II} to be associated in a way that its reactivity is limited. While these complexes may be with NOM, there is also evidence to suggest that the Hg may also be associated with metal-sulfide entities (Hsu-Kim and Sedlak, 2005), that have been termed "metal sulfide clusters" or "colloidal metal sulfides" in the literature. Thus, there is sufficient evidence to suggest that Hg^{II} complexation in WWTP effluents is not controlled by the presence of inorganic ligands such as Cl^- and OH^-, but that the Hg^{II} is complexed to strong binding ligands that are either NOM or metal-sulfide colloids, which form part of the dissolved fraction as a result of typical filtration approaches (0.4 μm filters) used in collecting samples.

There is evidence in the literature, especially in freshwater systems influenced by industrial effluents, that "inorganic sulfide" persists in these oxic environments (Rozan et al., 2000; 2002). Based on the measurement techniques and knowledge of the speciation of metals in the presence of inorganic sulfide, it is likely that the majority of this sulfide is bound up in metal complexes or clusters, and is therefore not truly free sulfide in solution (Rozan et al., 2000; 2002; Rearick, 2004). The presence of these metal complexes appears to stabilize the sulfide against rapid oxidation and thus they can persist in solution for a substantial period (Rozan et al., 2000; Hsu-Kim and Sedlak, 2005). Given the strong binding of Hg with sulfide, such complexes with Hg would be expected to persist in environmental waters, and preliminary studies suggest this is the case (Hsu-Kim and Sedlak, 2005). Similar strong ligands have been found with silver (Ag) (Herrin et al., 2001; Smith et al., 2004) and copper (Cu) (Van Veen et al., 2002) in WWTP effluent, and these complexes have been shown to be stable for periods greater than 24 hours. Given the similar chemistry of Ag^{I} and Hg^{II}, it can be expected that Hg complexes would behave in a similar fashion. The data of Hsu-Kim and Sedlak (2005) suggest these types of complexes can be important in WWTP effluents. In the presence of sulfide, Hg-sulfide complexes will be the dominant ligand complexing Hg in solution, based on chemical thermodynamic speciation modeling (Miller 2005; Benoit et al., 2001).

However, it has also been shown recently that the organic complexes of Hg and MeHg, especially those with NOM (Figure 2-2), cannot be ignored in environmental media (e.g. Ravichandran 2004) and in many situations these complexes are likely to dominate the speciation of both Hg and MeHg (Lamborg et al., 2003 and references therein; Benoit et al., 2001; Miller et al., 2007). This is also potentially true for WWTP effluents (Hsu and Sedlak, 2003; Hsu-Kim and Sedlak, 2005). In terms of bioavailability, it is often the "first-cut" conclusion to ignore the organic complexes as non-bioavailable as they are large and have an overall negative charge due to the overall negative charge present on large organic ligands in aqueous solution. Note that in some investigations, the interaction of Hg and NOM has been modeled as a 1:1 interaction. While this may be so, the complex formed is not positively charged, as is often indicated, as the NOM has multiple sites for binding and Hg prefers to form two associations (e.g. Skyllberg et al., 2006). Therefore, one can consider this interaction as the formation of multiple bonds of Hg with one NOM molecule. Thus the overall charge of this complex is due to that of the surface charge of the organic matter (Morel and Hering, 1993), which is likely negatively charged except at low pH - lower than typical environmental conditions. Indeed, it is not clear whether it is better to model and understand the interaction of Hg and MeHg with NOM as that of a true complex in solution or that this should be dealt with under the auspices of an interaction between Hg and MeHg and colloidal matter, taking into account the "surface charge" on the organic molecule.

While the literature on this subject has increased markedly in recent years, there are still unresolved complications. One aspect of recent study is the interaction between NOM and Hg in the presence of sulfide and the demonstration that Hg-S complexes appear to interact with NOM (Miller et al., 2007; Hsu-Kim and Sedlak, 2005). These complexes formed via a Hg-S-NOM interaction are also likely not readily bioavailable as the kinetics of Hg release or exchange for such molecules/entities is likely slow compared to that of the inorganic complexes (Lamborg et al., 2003). Thus, for now, the research team suggests that Hg-NOM complexes, and especially HgS-NOM entities, are not readily bioavailable because of kinetic effects. This again is a matter of some debate.

It has been proposed that two interactions could account for the strong binding of Hg in complexes in WWTP effluents that are resistant to degradation or chemical exchange reactions. The compounds/complexes themselves have not been isolated but there is some information available on their chemistry and reactivity. One hypothesis is that these compounds are Hg-S clusters i.e. Hg_xS_y, where x, y > 1. Such clusters have been shown to exist for Zn, with x and y varying from 2 to 8 (Rozan et al., 2000). The Zn clusters have been shown to be resistant to decomposition in oxic waters, and it is likely that similar clusters with Hg would be similarly resistant (Hsu-Kim et al., 2005). It is not clear whether the Hg clusters are of the form Hg_xS_y, or whether they are Hg-polysulfide entities (HgS_y, y = 3-6) (Jay et al., 2000). The bioavailability of such compounds is not known, but these compounds would be neutral and likely hydrophobic and so may partition into membranes passively, although their size may limit their rate of uptake. For example, Jay et al. (2002) estimated that the uptake of HgS_y by sulfate-reducing bacteria would be 1-2 orders of magnitude slower than HgS^0 and demonstrated that in the presence of sulfide, uptake of the polysulfide complexes was not important (as measured by differences in Hg methylation rates) in terms of Hg methylation. It is not possible to directly apply these results to WWTP effluents but these results do suggest that the bioavailability of the proposed Hg-S cluster entities may be lower than that of other forms of Hg in solution.

The other hypothesis is that the strong Hg ligands in solution are those involving NOM. As the complexes of Hg with NOM are relatively labile, and can be shown to be exchangeable in the presence of other strong Hg binding ligands, it is likely that the strong complexes do not involve only Hg and NOM. Miller et al. (2007) have shown that the speciation of Hg in the presence of NOM and sulfide is not comparable to thermodynamic prediction based on current known Hg complexes. They suggest an interaction where either the neutral Hg-S complexes (HgS^0 and $Hg(SH)_2^0$) hydrophobically partition into the "organic matrix" of the NOM or form a tertiary complex such as HS-Hg-SR, where the Hg is bound to an inorganic sulfide ligand (HS^-) and a thiol ligand of the NOM (designated here as RS^- where R is the remainder of the NOM molecule). Either of these types of compounds would likely be relatively unavailable for uptake into microorganisms, and both forms are also likely resistant to degradation. Currently, there is insufficient evidence to determine whether the strongly bound Hg in WWTP effluent is primarily inorganic or is some inorganic-organic matter association. This is obviously an area for future investigation and research.

2.4.2 Organic Mercury Speciation and Bioavailability

In terms of the organic Hg compounds, all evidence suggests that for WWTP effluent the most important organic Hg compounds are the complexes of CH_3Hg^{II}. There is no evidence in the literature for $(CH_3)_2Hg$ in such waters, or for larger organic Hg compounds, such as phenylmercury. Also, in most WWTP effluents it appears that %MeHg is <10% (e.g. Balcom et al., 2004). Therefore, the focus is on CH_3Hg.

As with Hg^{II}, CH_3Hg^{II} forms a variety of complexes with inorganic and organic ligands (Figure 2-2). The primary difference between the two forms of Hg in terms of their complexation is that the Hg free ion has a +2 charge (Hg^{2+}) while the CH_3Hg ion has a +1 charge (CH_3Hg^{+}). In most aquatic environments, the concentration of Hg^{2+} is very small and a minor constituent of the system, based on equilibrium modeling studies (Mason et al., 1996; Morel and Hering 1993), and therefore its presence can be ignored in terms of bioavailability. The same is not true for CH_3Hg^{+}, which can exist in solution as a relatively high fraction of the total CH_3Hg^{II} (>1%) under a variety of conditions (Figure 2-3). In addition, Hg^{II} tends to form primarily linear complexes with two ligands, although 1:1 associations are found, as are higher order complexes, such as $HgCl_3^-$ and $HgCl_4^{2-}$. For CH_3Hg^{II}, most complexes are 1:1 (Figure 2-2).

2.4.3 Mercury in the Particulate Phase

One final consideration is Hg attached to or incorporated into particulate material, as this can be the dominant fraction for Hg^{II} in WWTP effluent. Here particles are defined as those retained on a filter, mostly >0.4 μm because of the primary filtration cutoff used by most

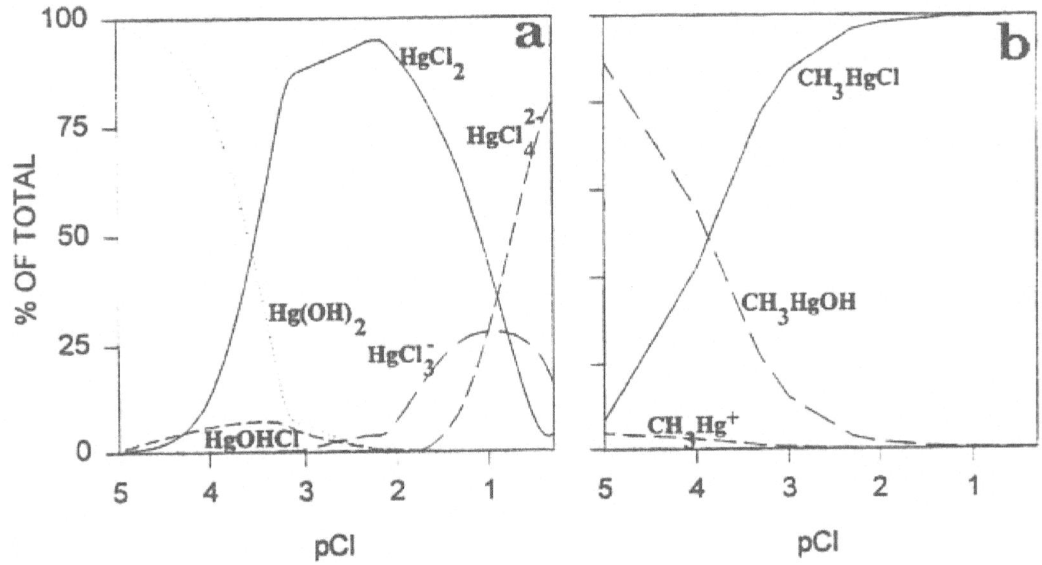

Figure 2-3. The Speciation of Mercury and Methylmercury at Different Chloride Concentrations at pH 6. Reprinted with permission from Mason et al., *Environmental Science & Technology*, American Chemical Society (1996).

investigators. For particles, either Hg or CH_3Hg bound to the surface of a mineral particle or the organic coating on such a particle could be potentially bioavailable, so there is the need to consider the form and speciation of Hg in the particulate phase. If Hg is incorporated into a mineral such as cinnabar (HgS^0(s)), the concentration of the Hg in solution in equilibrium with the solid phase determines the bioavailability. Therefore, while the solid itself may not be bioavailable, it can be a source of bioavailable Hg in solution. However, there is little evidence for the existence of HgS minerals except in specific locations unrelated to WWTPs and in most environmental situations. It is more likely that the Hg is bound to the surface or, or incorporated into, iron (Fe) sulfide solids due to the much higher abundance of Fe in these systems and its tendency to precipitate at lower sulfide levels. Modeling studies and field investigations of Hg in porewater show that the dissolved concentration does not reach the level necessary to precipitate the Hg-S solids (Benoit et al., 1999a) except under unusual conditions. However, both Hg and MeHg bind strongly to amorphous FeS. Therefore in the presence of this solid, most of the Hg and MeHg will be bound to the solid phase (Miller, 2005).

In oxic waters, Hg forms relatively weak bonds with the surface of iron oxides and other metal oxides (Miller, 2005) – so if Hg were in this form in the particulate phase, it would be potentially bioavailable. However, various studies suggest that under most environmental conditions, and likely in the effluent of WWTPs, there is sufficient organic matter in solution and in the solid phase (i.e. surface organic coatings on particles) that the binding of Hg to these organic phases controls its partitioning, or that both Hg-NOM and Hg-S complexes exist (Miller et al., 2007; Hsu-Kim and Sedlak, 2005). Under such conditions, the Hg is much less available for bioaccumulation, and would likely remain so even if the oxide phases were reduced due to changes in redox status. Similarly, oxidation of the Fe-sulfide phases would likely result in the transfer of the Hg from the sulfide matrix to an organic matrix resulting in little change in bioavailability and little release to the water column, as was found in a study in the Hudson River (Heyes et al., 2004), and in mesocosm experiments (Kim et al., 2006).

2.4.4 Mercury Uptake by Microorganisms

It is thought that the primary forms of Hg and MeHg assimilated by microorganisms are dissolved species, as complexes of either Hg^{II} or CH_3Hg^{II}. While both Hg^0 and $(CH_3)_2Hg$ have relatively large octanol-water partition coefficients (K_{ow}) (Mason et al., 1996), a measure of the propensity of a compound to partition into hydrophobic substances such as a biological membrane (Schwarzenbach et al., 1993), it is not only an ability to partition that is important in bioavailability and bioaccumulation - there needs to be a sink for the accumulated substance inside the cell to maintain the concentration gradient between the inside (i.e. within the cell) and outside of the membrane (Mason et al., 1996; Jay et al., 2002). In the case of Hg and MeHg, which are normally accumulated in tissue to levels many orders of magnitude above their concentration in the surrounding environment, this is only possible if there is a "sink" for the accumulated Hg within the biological tissue, especially for microorganisms. Mason et al. (1996) showed that Hg^0 and $(CH_3)_2Hg$ were not accumulated into phytoplankton cells. The reason for this was the lack of binding or transformation of the Hg^0 and $(CH_3)_2Hg$ inside the cell. Therefore, for the purposes of this study focusing on aquatic systems, it was concluded that Hg^0 and $(CH_3)_2Hg$ are not bioavailable unless Hg^0 is oxidized and converted into ionic Hg^{II} within the tissue, and $(CH_3)_2Hg$ is degraded to CH_3Hg^{II} complexes.

There is an increasing body of research demonstrating the ability of the neutrally-charged complexes of both Hg and MeHg to cross biological membranes by passive diffusion. For example, Mason et al. (1995; 1996) showed this for the uptake of the neutral chloride complexes – $HgCl_2^0$ and CH_3HgCl^0 – by phytoplankton, while Benoit et al. (1999a; 1999b) showed that the same was true for the neutral complexes of Hg with sulfide – HgS^0 and $Hg(SH)_2^0$ – in terms of their bioavailability to methylating bacteria. Note that while the complex HgS^0 is written as such in many papers, it is likely that this complex is HOHgSH, as Hg prefers to form linear complexes (Tossell, 2001). These determinations were based on both uptake studies and laboratory determinations of hydrophobicity, as exemplified by the K_{ow}. These studies showed that the other species in solution, which were mostly negative charged complexes, were not bioaccumulated to any significant degree. Lawson and Mason (1998) performed studies in the presence of organic ligands and showed that these complexes are also taken up under the right conditions, albeit about an order of magnitude more slowly, than the small, neutrally-charged inorganic complexes. Indeed, there is evidence of slow uptake of Hg and MeHg under most conditions examined and this may be via a ligand exchange surface reaction. However, this has not been carefully examined in any published literature studies.

Since the initial studies of uptake of neutral complexes, estimation of solution speciation and estimation of uptake based upon this have become the basis for assessments of bioavailability in a number of situations. More recently, there has been evidence that positively charged complexes of ionic Hg can be taken up by microorganisms via some active pathway. The evidence for positively charge Hg^{II} complex uptake was first demonstrated in engineered solutions, such as those used for "bioreporter" studies (Golding et al., 2002), and have been further demonstrated in laboratory manipulations (Najera et al., 2005). However, the uptake of positively charged Hg^{II} complexes by microorganisms is not a major factor due to their low concentration in environmental solutions. The same may not be true for MeHg, which can exist in solution as CH_3Hg^+ at higher relative concentrations (>1%) (Figure 2-3). Overall, the speciation of Hg and MeHg in waters has a complex influence over the bioavailability of Hg and MeHg to microorganisms. This complexity needs to be considered in detail when examining bioavailability in WWTP effluents.

2.4.5 Bioavailability of Sediment Mercury to Benthic Organisms

There is the potential for Hg and MeHg to be assimilated from sediment or suspended material upon their digestion by invertebrates and other higher trophic levels organisms. While the Hg and MeHg may be within the sediment solid phase, their release into solution (in the gut) is required prior to digestion. This can occur during the digestion process, as the accepted paradigm of bioaccumulation during digestion is that only soluble compounds are absorbed across the gut lining. Studies of Hg, MeHg and a number of contaminants (Mason, 2002) indicate a strong relationship between the degree of solubilization of a metal by intestinal fluids during digestion and the overall bioaccumulation into the organism. The extent of release of Hg and MeHg from sediments is a function of an organism's intestinal fluid chemistry and overall physiology. The degree of bioaccumulation depends on the each organism's unique digestion processes as well as the chemistry of the sediments ingested (e.g. Chen and Mayer, 1999; McAloon and Mason, 2003). From a chemical perspective, the degree of solubilization of Hg and MeHg during digestion for most invertebrates appears to be a competitive ligand exchange process between the ligands in the sediment (particulate organic matter (POM) and acid volatile sulfide (AVS) phases) and the ligands in the intestinal fluids (primarily amino acids) (Mason, 2002). Thus the degree of solubilization from the same sediment by the intestinal fluids of different organisms is different and is related to amino acid content of the individual organism's intestinal fluid (Chen and Mayer, 1999; McAloon and Mason, 2003).

Overall, it is the strength of binding of the Hg^{II} and MeHg to the solid phase that determines their bioavailability to benthic organisms, and it has been shown that bioavailability and bioaccumulation is inversely related to NOM (Lawrence et al., 1999; Mason, 2002; McAloon and Mason, 2003). Thus, Hg^{II} and MeHg in sediments with a high NOM are much less bioavailable than sediments with low NOM. Additionally, while it is also true that sediments with a high NOM typically contain more Hg and MeHg than low NOM sediments, it can be concluded that overall the Hg^{II} and MeHg will be accumulated into benthic organisms to a lesser degree from high NOM sediments. Thus, it is possible to assess the relative bioavailability of Hg and MeHg in sediments and the suspended particulate if the bulk parameters, such as NOM of the material, are known.

CHAPTER 3.0

ENHANCERS AND MITIGATORS OF MERCURY BIOAVAILABILITY

3.1 Identification of Bioavailability Enhancers and Mitigators

The interaction of factors that may increase or decrease the bioavailability of Hg and MeHg in effluents and receiving waters is complex. It is difficult to be completely specific about these factors as in most cases there is high variability across systems and differences that result from the type of treatment plant, the type of receiving waters, and the extent of dilution of the effluent in the receiving waters. Much of the information needed for a thorough examination of the topic has only been recently collected, is available for only a small number of plants, or has been obtained with specific targeted research studies rather than from long-term monitoring data. To enable a detailed examination of the subject would require information for a number of different treatment plants for the following parameters:

- Suspended and biosolid Hg and MeHg concentration and solid characterization (e.g. organic and sulfur content);
- Filtered Hg and MeHg;
- Ancillary parameters such as total suspended solids (TSS), sulfate, chloride, dissolved organic matter (NOM), and overall ionic strength;
- Receiving water Hg, MeHg and the above ancillary parameters, including oxygen demand of the system; and
- Characterization of the associations of Hg and MeHg in filtered effluent water (i.e. inorganically or organically-complexed; extent of colloidally-associated Hg and MeHg).

Even so, there is a rapidly growing body of knowledge concerned with the factors that affect mercury methylation and bioaccumulation in natural waters. Table 3-1 below provides a brief summary of the most important factors enhancing or mitigating the bioavailability of T-Hg and MeHg in effluents. The relative importance of each parameter is indicated as high, medium, low, no effect (H, M, L, N). In those cases where the issue is too complex for a particular impact to be estimated with certainty, the impact is indicated as uncertain (U). In each case, when considering the impact of a particular parameter, it is assumed that changes in this parameter occur without any concomitant changes in the other factors (i.e. all else remains equal). The details of how these conclusions were reached are dealt with in depth in the subsequent discussion.

Table 3-1. Identification of Enhancers/Mitigators of Bioavailability in Effluents and Receiving Waters and the Magnitude of their Potential Effects.

Enhancer/	Predicted influence of enhancer/mitigator on bioavailability of Hg				
	Mechanisms increasing availability	**Importance***		**Mechanisms decreasing bioavailability**	**Importance***
Effluent and Plant					
T-Hg Concentration	Higher T-Hg would lead to increased bioavailable inorganic Hg (3.2.1)	H			
MeHg Concentration	Higher MeHg is postulated to reflect higher bioavailability (3.2.2)	H			
TSS Concentration				Higher TSS would likely reduce bioavailability due to sorption. (3.2.1)	H
Strength of partitioning of T-Hg and MeHg to solids				Stronger binding of T-Hg and MeHg to solids reduces bioavailability to microbes (decreases dissolved concentrations) and to higher trophic organisms (by decreasing the extent of solubilization during solid digestions). Also, decreases T-Hg bioavailability for conversion to MeHg, and MeHg availability for demethylation. (3.2.3)	M
Organic content of effluent	Higher organic content of effluent may stimulate sulfate reduction by providing an available carbon source.	M		Higher organic content in solution would decrease bioavailability due to strong complexation. (3.2.6)	H
Presence and strength of Hg-ligand complexes				Presence of these complexes (e.g. Hg-S-NOM) would decrease bioavailability (3.2.4)	H
Presence of major inorganic ions (e.g. sulfate, sulfide, chloride)	No evidence suggesting impact, sulfate is not limiting, Cl appears too low to have a complexing effect (3.2.4). However, the presence of sulfide creates opportunities for formation of bioavailable HgS^0 forms.	N (SO_4^{2-}, Cl^-) M (S^{2-})			
Receiving Waters					
Dilution ratio	May increase or decrease depending on chemistry of receiving waters (3.3.1)	U			

Table 3-1. Identification of Enhancers/Mitigators of Bioavailability in Effluents and Receiving Waters and the Magnitude of their Potential Effects (continued).

Enhancer/ Mitigator	Predicted influence of enhancer/mitigator on bioavailability of Hg				
	Mechanisms increasing availability	Importance*		Mechanisms decreasing bioavailability	Importance*
Receiving Waters					
Ionic Strength (Conductivity)				Higher ionic strength tends to coagulate particles resulting in more rapid sedimentation, thereby decreasing bioavailability in the water column. Methylation and release from sediments could mitigate this effect. (3.3.2)	L
Relative concentration of NOM, TSS and POC				If relatively high, these could diminish the bioavailability of Hg or MeHg added via effluent. Unlikely to be lower than in effluent. High TSS and POC may not decrease bioavailability of MeHg to higher trophic level organisms. (3.3.1)	L
Low DO / high BOD	Low oxygen environments and high biological oxygen demand could lead to enhancement of methylation in the receiving waters. (3.3.1)	H			
Sulfate concentration	Increasing sulfate in receiving waters may increase methylation potential. (3.3.1)	M			
pH	For low pH receiving waters, methylation could be enhanced (3.3.3)	M			
Receiving water geometry	Shallow, slow flowing systems will enhance the transfer of Hg to sediments. (3.3.3)	L - M			
Trophic Status				Increased growth rate and biomass leads to lower Hg concentrations in biota (biodilution) (3.3.3)	M
Ecological Sensitivity	Some natural waters seem to be susceptible to enhanced bioaccumulation of methylmercury. Lakes>streams>estuaries Oligotrophic>dystrophic> eutrophic. (3.3.3)	H			
Notes: * Entries are qualitative (H = high, M = moderate, L= low, N = no effect, U =uncertain or case specific. Numbers in parentheses refer to the section in which the enhancing/mitigating factor is discussed. Numbers in parentheses refer to the chapter section in which the factor is discussed.					

It is of interest to compare the factors identified above to those identified by Moore et al. (2003). In their report on the development of ambient water quality criteria for mercury, they found pH, DOC and MeHg concentration in water to be the most important predictive factors for mercury in fish tissue. Using step-wise multiple linear regression, they found pH to be the most important factor, with DOC and MeHg concentration being less important but still statistically significant. Other correlations were also found between fish tissue concentration and total mercury in water, color, conductivity, and sulfate. While these factors were identified empirically, on the whole, the results are consistent with the factors identified in Table 3-1 above from biogeochemical principles.

3.2 Factors Related to the Effluent

3.2.1 Total Mercury Concentration

In this report, most of the discussion of bioavailability focuses on relative quantities; that is, the amount of Hg and MeHg in the effluent and receiving waters that is bioavailable relative to the total mercury concentration. However, it is obvious that the total concentration of Hg and/or MeHg in the effluent is an important variable and that all discussions, while focusing on relative availability, acknowledge the importance of concentration. This would not be the case if the total Hg concentration in effluents was constant or similar across systems, but, as shown in the data compilation in Table 3-2, average total Hg concentration of effluent can vary by at least two orders of magnitude across WWTPs. For example, the limited measurements of the outfall from Clean Waters of New York (n=2) indicate a total Hg concentration of <1 ng/L, certainly less or equivalent to that of the receiving waters. Similarly, the Oakwood Beach facility has a low average concentration (1.25 ng/L). In contrast, within the same watershed, there are a number of plants with effluent concentrations > 20 ng/L, on average, with individual measurements being over 100 ng/L. The same can be seen in the data from other regions of the US. While some of these differences may reflect historical changes in effluent concentrations over time, there are facilities where concentrations > 100 ng/L have been recently measured. It is probable that there has been a decrease in total Hg in effluent resulting from various activities to minimize mercury and the requirement to make measurements using "cleaner" sampling and more sensitive analytical techniques. However, there is still a large variability in total Hg concentration across systems.

Design and operation of the WWTP has an impact on the effluent Hg concentration. For example, if incineration is used and the scrubber water is recycled to the effluent steam, higher effluent concentrations will likely result. This was indeed the case for the facility studied by Balogh and Liang (1995). Another obvious example was the plant studied by Gilmour and Bloom (1995) where liquid Hg seals were used in the trickling filters. Again, this design led to a high effluent Hg and MeHg concentration. Overall, the degree of recycling of the water within the plant may be a factor leading to higher Hg and possibility higher MeHg in the effluent. This is discussed further below. Currently, there is not a sufficient database on Hg concentrations in effluent for a variety of plant types and operations to allow a more detailed examination of the relationship between plant type and effluent Hg, but this is an area which should be examined in more detail as more information becomes available.

Table 3-2. A Compilation of the Average Concentrations of Total Hg, and Its Relative Fractionation between Particulate and Solid Phases, for a Number of WWTPs in the U.S.

Location	Region	Sample Period	Avg. T-Hg ng/L	Range ng/L	Filtered T-Hg ng/L	Particulate T-Hg mg/kg	TSS Range mg/L	Ref.
Mattahasset-Cromwell	Long Island Sound	1996-97	20	6 to 60	0.4 to 54	1.3 (0.1 to 5.2)	14.3	1
Hartford	Long Island Sound	1996-97	13	5 to 20	1 to 16	1 (0.1 to 3)	8.3	1
Greenwich	Long Island Sound	1996-97	5.4			6.7	6.9	1
New Haven	Long Island Sound	1996-97	13.1			2.5	5.5	1
Norwalk	Long Island Sound	1996-97	4			7	3	1
Stamford	Long Island Sound	1996-97	26			5.4	37.6	1
Sacramento	San Francisco delta	2006	8.1		3.2			2
San Jose/Santa Clara	San Francisco delta	Recent	2	1 to 5	1.4	1 to 2	2 to 5	3
WWTP 1	California	2001	8.6		5.8			4
WWTP 2	California	2002	25.2		39.7			4
WWTP 3	California	2002	52		40.4			4
Various	Winnepeg Overall	1994/95		3 to 14				5
St Paul	Minnesota	90's	287	87 to 630				6
Other WI & MN	Wisconsin, Minnesota	<1990		125 to 820				6
Hubbing South	St Louis River	<1995		2 to 18				7
Onondaga Lake	New York	1995/96	32	2 to 70				8
Newtown Creek	New York/New Jersey		39.1	20 to 65	10.06			9
Bowery Bay	New York/New Jersey		13.5	5 to 20	7.4			9
Port Richmond	New York/New Jersey		10.24	4 to 15	3.8			9
Oakwood Beach	New York/New Jersey		1.25	1 to 5	3.2			9
Owls Head	New York/New Jersey		21.7		0			9
Rockaway	New York/New Jersey		18.7	6 & 32	3.6			9
Clean Waters of NY	New York/New Jersey		0.54	<1	0.32			9
Tallman Island	New York/New Jersey		9.08	3 to 22	1.02			9
Coney Island	New York/New Jersey		6.58	-	2.94			9
North River	New York/New Jersey		4.26	-	12.3			9
Rockland County	New York/New Jersey		64	30 to 110	15.18			9
Yonkers	New York/New Jersey		20.7	9 to 50	6.38			9
Poughkeepsie (C)	New York/New Jersey		46.1	25 to 90	4.66			9
26th Ward	New York/New Jersey		12.1	9 & 15	2.04			9
Red Hook	New York/New Jersey		9.02	<10	2.5			9
Wards Island	New York/New Jersey		7.94	6 to 11	2.74			9
Hunts Point	New York/New Jersey		9.68	8 to 10	3.34			9
Fresh Kills Landfill	New York/New Jersey		55.6	13 & 20	84.3			9
Rensselaer	New York/New Jersey		14.04	-	3.46			9
Jamaica	New York/New Jersey		14	13 to 15	2.72			9

Sources: 1= Balcom et al. (2004); 2 = Sacramento Regional County Sanitation District Bioaccumulation Study Workplan (2005); 3 = ESD (2007); 4 = Hsu (2004); 5 = Bodaly et al. (1998); 6 = Balogh and Liang (1995); 7 = St Louis River TMDL Report; 8 = Driscoll et al., 1997; 9 = CARP Data on WWTPs in New York/New Jersey Harbor.

Another major factor is (total) suspended solids (TSS) concentration, as inorganic Hg and to a lesser degree MeHg, are strongly associated with particles in the effluent stream. This is shown for example in data compiled by Hsu (2004) (Figure 3-1) which shows a strong relationship between TSS and T-Hg concentration. While there is clearly variability, the data demonstrates an overall a relatively strong correlation. A similar correlation is shown in Figure 3-2a for WWTPs discharging into Long Island Sound (Balcom et al., 2004), although the relationship is almost entirely driven by the two high data points and without these data the relationship is not significant. A similar weak relationship was found for the San Jose/Santa Clara (SJ/SC) plant (Figure 3-3) (ESD, 2007). Thus, while the relationship of TSS and total mercury may vary from plant to plant, it is clearly important to consider its impact in terms of bioavailability of the effluent. This issue is discussed in greater detail below in Section 3.2.3.

As noted below, there are many factors that can influence the bioavailability of Hg and MeHg in effluents and natural waters. Some of these factors, like the concentration of dissolved natural organic matter (NOM), have a complex role. Increasing NOM could possibly will lead to increased Hg concentration as NOM helps "keep Hg in solution". However, the presence of NOM also reduces Hg bioavailability. So, increasing Hg concentration in solution as a result of higher NOM levels will not result in higher relative bioavailability, but may result in more bioavailable Hg if the total Hg concentration is higher. This has been shown in the literature in many instances. All else being equal, therefore, increasing the total Hg concentration in effluent should lead to a higher amount of bioavailable *inorganic* Hg. However, the factors that influence total Hg concentration in solution may not be the same as those that enhance Hg conversion to MeHg, so a higher total Hg concentration cannot be taken as indicative of a higher MeHg concentration. Indeed, in many relatively contaminated systems it has been shown that the relative MeHg concentration (i.e. % MeHg) is lower than in more pristine environments (Benoit et al., 2003). Thus, the impact of increasing total Hg concentration on bioavailable inorganic Hg is relatively straightforward, but it cannot be stated with any certainty that increasing total Hg concentration will result in higher MeHg concentration, as many other factors besides total Hg concentration influence the relative relationship between total Hg and MeHg.

3.2.2 Relative Methylmercury Concentration and Methylation Potential

The relative concentration of methylmercury to total mercury is important as it indicates not only the presence of bioavailable inorganic mercury, but also of the existence of conditions to convert bioavailable inorganic Hg to MeHg. A review of the available data on the fraction of total Hg as MeHg (relative methylmercury or %MeHg) in the effluent stream shows that it is not constant across facilities. As with total mercury, there may be processes within the plant enhancing MeHg loading at the outfall.

The concentration of MeHg in effluents range from low values (~0.01 ng/L) to values above 2 ng/L for some facilities (Table 3-2). High average values of 1.7 and 0.94 ng/L were reported for two facilities in California, for example. The average value for all the data compiled in Table 3-2 is 0.3 ng/L. Relative MeHg, which is based on less data, ranges from <1% to 9%, with an average of 2%. This range in relative MeHg is similar to that of receiving waters, which suggests that the fraction as MeHg is not widely different in effluents than other waters. Thus, there should not be a major enhancement of the relative MeHg in the receiving waters except under conditions of a high %MeHg in the effluent and a high ratio of effluent flow to receiving water flow.

Table 3-3. A Compilation of the Average Concentrations of Methylmercury, and its Relative Fractionation between Particulate and Solid Phases, for a Number of WWTPs in the U.S.

Location	Region	Avg. MeHg ng/L	Range ng/L	Filtered MeHg ng/L	Particulate MeHg ng/g	Avg. %MeHg	Ref.
Mattahasset-Cromwell	Long Island Sound	0.42	0.1 to 1.2	0.05 to 0.7	5 to 20	2.1	1
Hartford	Long Island Sound	0.3	0.2 to 0.7	0.04 to 0.6	5 to 40	2.3	1
Greenwich	Long Island Sound	0.17				3.1	1
New Haven	Long Island Sound	0.14				1.1	1
Norwalk	Long Island Sound	<0.1				<2.5	1
Stamford	Long Island Sound	0.46		0.28		1.8	1
Sacramento	San Francisco delta	0.73	up to 3	0.29		9.0	2
San Jose/Santa Clara	San Francisco delta	1.4					3
Various	Winnepeg Overall		0.13 to 0.56				4
Onondaga Lake	New York	1.5	0.5 to 3.9			4.7	5
Newtown Creek	New York/New Jersey	*0.62**		0.37		*1.6*	6
Bowery Bay	New York/New Jersey						6
Port Richmond	New York/New Jersey	*0.03*		0.017		*0.3*	6
Oakwood Beach	New York/New Jersey	*0.09*		0.053		*7.1*	6
Owls Head	New York/New Jersey						6
Rockaway	New York/New Jersey	*0.10*		0.06		*0.5*	6
Clean Waters of NY	New York/New Jersey	*0.01*		0.006		*1.9*	6
Tallman Island	New York/New Jersey	*0.07*		0.04		*0.7*	6
Coney Island	New York/New Jersey	*0.13*		0.08		*2.0*	6
North River	New York/New Jersey	*0.20*		0.12		*4.7*	6
Rockland County	New York/New Jersey	*0.87*		0.52		*1.4*	6
Yonkers	New York/New Jersey	*0.20*		0.12		*1.0*	6
Poughkeepsie (C)	New York/New Jersey	*0.63*		0.38		*1.4*	6
26th Ward	New York/New Jersey	*0.10*		0.06		*0.8*	6
Red Hook	New York/New Jersey	*0.17*		0.1		*1.8*	6
Wards Island	New York/New Jersey	*0.03*		0.02		*0.4*	6
Hunts Point	New York/New Jersey	*0.07*		0.04		*0.7*	6
Fresh Kills Landfill	New York/New Jersey	*0.40*		0.24		*0.7*	6
Rensselaer	New York/New Jersey	*0.10*		0.06		*0.7*	6
Jamaica	New York/New Jersey	*0.03*		0.018		*0.2*	6
Brentwood	San Francisco delta	<0.02	< 0.02				7
Deuel Voc. Inst	San Francisco delta	<0.02	< 0.02				7
Discovery Bay	San Francisco delta	0.2	0.025 to 2.0				7
Lodi White slough	San Francisco delta	0.13	0.02 to 1.2				7
Manteca	San Francisco delta	0.22	0.04 to 0.36				7
Rio Vista	San Francisco delta	0.16	0.04 to 0.52				7
Rio Vista Trilogy	San Francisco delta	-					7
Flag City	San Francisco delta	0.09	0.025 to 0.15				7
Walnut Grove	San Francisco delta	1.69	0.14 to 2.9				7
Stockton	San Francisco delta	0.94	0.76 to 3.4				7
Tracy	San Francisco delta	0.15	0.02 to 0.42				7
West Sacramento	San Francisco delta	0.05	0.02 to 0.085				7

Data taken from the literature. 1= Balcom et al. (2004); 2 = Sacramento Regional County Sanitation District Bioaccumulation Study Workplan (2005); 3 =San Jose/Santa Clara Water Pollution Control Facility Control Plant Mercury Fate and Transport Study; 4 = Bodaly et al. (1998); 5 = Driscoll et al., 1997; 6 = CARP Data on WWTP in New York/New Jersey harbor; 7 = Delta TMDL Draft methylmercury report (Abu-Saba et al).

* Values in italics are estimated as only the dissolved methylmercury concentration was measured for these plants.

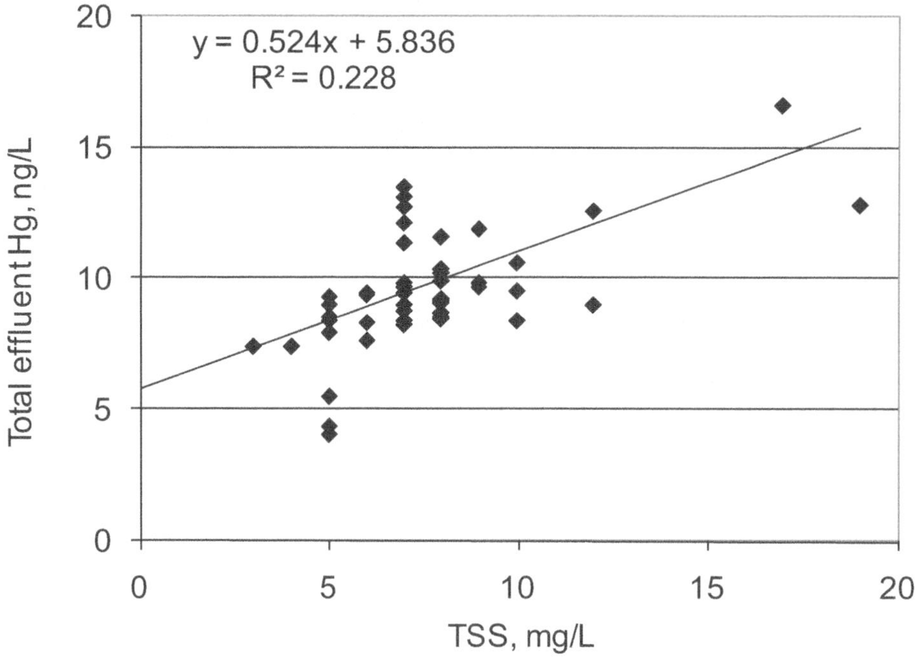

Figure 3-1. Relationship between Total Mercury and Total Suspended Solids in the Final Effluent of a Wastewater Treatment Plant. Reprinted with permission from Hsu. H, *Mercury Biogeochemistry and removal during municipal wastewater treatment* (2004).

While it is not clear if MeHg is produced within WWTPs, there is indication that this may sometimes occur. Only three studies the researchers were aware of have sufficient information to allow such an assessment – the SJ/SC study, the Sacramento County Regional Sanitation District (SCRSD) MeHg study (Parmer et al., 2005), and the studies of Onondaga Metro (OM) (McAlear 1996).

In the SJ/SC facility, relative MeHg is near 1% in the raw sewage coming into SJ/SC, 1.6% in the primary effluent, <1% after the secondary treatment, 1.8% in the tertiary filter effluent and 2% in the final effluent. The latter increase is mainly because the filter backwash appears to have a high value of 5.8% MeHg. Why the backwash water has a higher fraction as MeHg is not known, but could be due to the recycled nature of this water. Overall, however, the data from the SJ/SC study show a steady decrease in MeHg concentration across the plant (1.57 ng/L in raw sewage; 0.04 ng/L in final effluent) with no clear indication of any significant production within the facility, even in the locations where production may be expected (i.e. associated with anaerobic treatment processes). This suggests strong removal of MeHg across

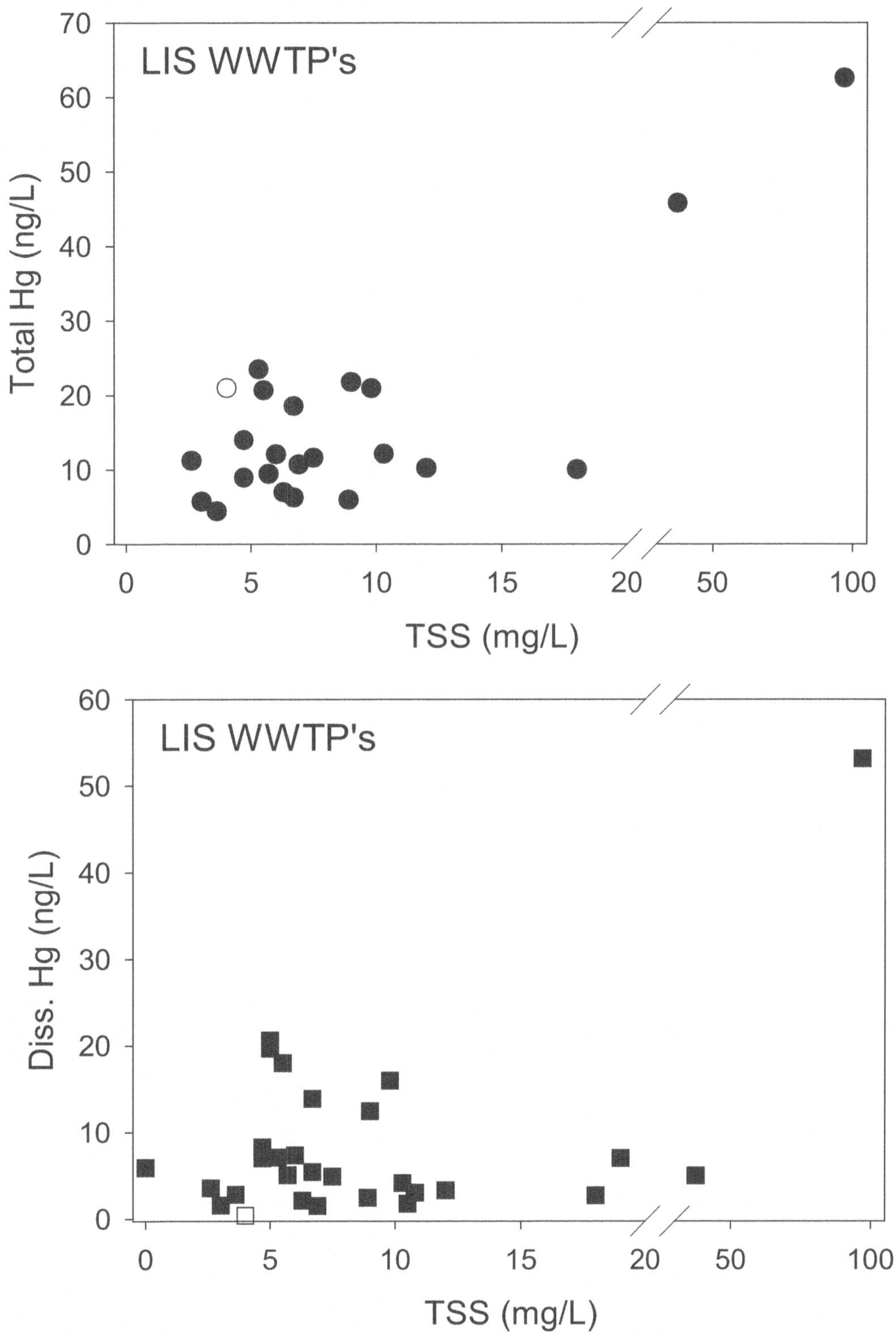

Figure 3-2. a) Relationship between Total Mercury and Total Suspended Solids for the Final Effluent of Treatment Plants Discharging into Long Island Sound. b) Relationship for Dissolved Hg and Total Suspended Solids. Data Source: Balcom et al. (2004).

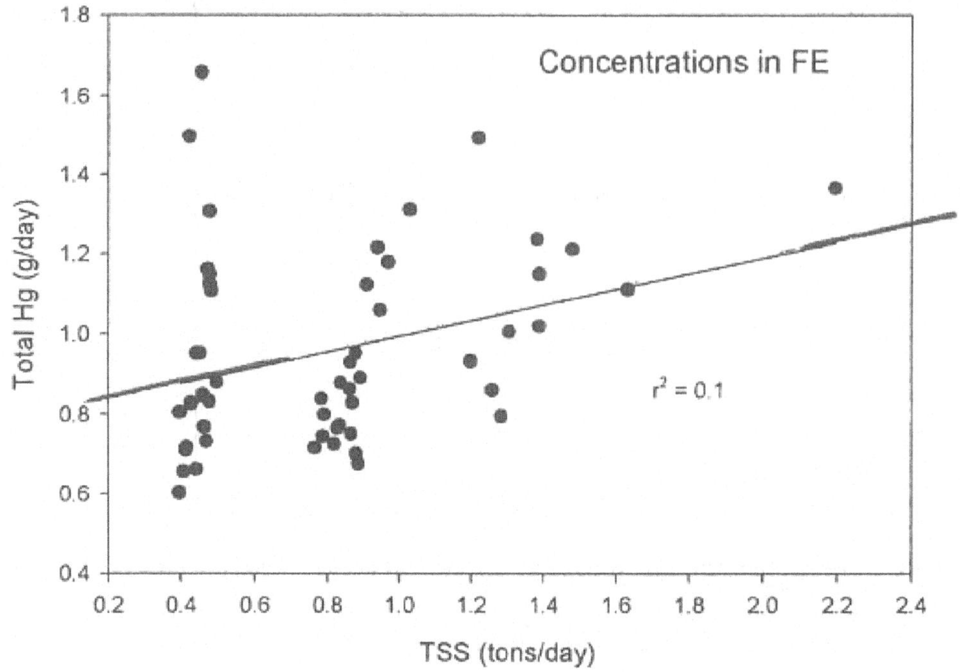

Figure 3-3. Relationship between Total Mercury and Total Suspended Solids in the Final Effluent of the San Jose/Santa Clara Wastewater Treatment Plant. Reprinted with permission from Environmental Services Department (2007).

the plant. At this facility, it is estimated that 90% or greater of the MeHg is removed, a high percentage and similar to that for total Hg.

For the OM plant, there is also overall a high removal of MeHg across the plant. Relative MeHg increased during secondary treatment to a substantial degree, although there is high variability, and is also high in the tertiary and final effluent compared to the influent waters (Figure 3-4b). In addition, MeHg after secondary treatment was higher in the summer than in winter. These differences are suggested by McAlear (1996) as production of MeHg within the plant and obviously contrasts the results from the SJ/SC facility where in situ methylation, if occurring, was small. Clearly, there appears to be something in the OM process that may be enhancing methylation within the plant and it appears that this may be related to seasonal effects, which would be consistent with higher bacterial activity in the summer. More investigation is required to ascertain the potential for in-plant methylation. It should be noted that the results for the OM plant are taken from a MS Thesis and these results have never been published in the peer-reviewed literature.

The Sacramento Regional County Sanitation District performed a study to investigate the fate and transport of MeHg throughout its treatment process and recycle streams. This study was conducted from September 2004 through May 2005 at the Sacramento Regional Wastewater Treatment Plant (SRWTP). The SRWTP uses conventional primary treatment followed by high purity oxygen secondary treatment, chlorine disinfection, and discharges treated effluent to the Sacramento River in the town of Freeport, south of the city of Sacramento.

The study utilized a two-phased approach; Phase 1Aand Phase 1B. Phase 1A focused on the principal liquid train locations and return flows. Phase1B sampling plan was developed from the results of Phase 1A and focused on data needed to investigate MeHg production and removal mechanisms in more detail. The study's main objectives were to evaluate the fate and transport throughout the plant, and to identify key factors that affect MeHg. Results concluded that approximately 75% of MeHg in the influent was removed by the treatment process, with the majority or the removal occurring in the secondary treatment process. Approximately 98% of total mercury (T-Hg) was removed by the treatment process. The difference in fractional removal for total Hg and MeHg would imply a net increase in %MeHg across the plant even if no net in situ methylation is occurring. However, the authors suggest that methylation may be occurring in the anaerobic digesters, as is evident by higher concentrations of MeHg in particles in the digesters and solids storage basins. Return flows from the onsite Biosolids Recycling Facility and Solids Storage Basins increase the influent MeHg loading by approximately 13%.

Known factors that enhance methylation are 1) sulfate concentration, but, as discussed below, this is not likely a limiting factor in WWTPs; 2) the presence of organic matter suitable for microbial degradation – also unlikely to be limiting in a WWTP; 3) the presence of the sulfate-reducing bacteria – again, these are likely present; and 4) the bioavailability of the Hg for the methylating organisms, which may be a factor in comparing the two plants. Another important factor is the degree of demethylation as the %MeHg is a result of the net effect of methylation and demethylation. It is possible that some process or condition in the SJ/SC plant design may enhance demethylation. This notion should be further investigated.

As mentioned previously, plant design and operation may also impact the extent to which in-plant methylation of Hg occurs. It is sensible that methylation would take place in areas where anaerobic conditions are encouraged. Two likely candidates come to mind, 1) in certain biological nutrient removal (BNR) processes where anaerobic conditions are developed either to enhance biological removal of phosphorus or to promote denitrification and volatile loss of nitrogen gas, or 2) in anaerobic digestion of sludges.

Different configurations for anaerobic digestion are possible within a WWTP plant. Such digesters can be single-stage or multi-stage, and can have different combinations of treatment parameters such as holding time, temperature, number of stages (U.S. EPA, 2006). These digesters are designed to enhance degradation of organic matter and to stimulate methanogenesis as the primary process for the final degradation step. It is known that methanogens are not important Hg methylators (e.g. Benoit et al., 2003) and so enhancement of methanogenesis should not lead to an enhancement of methylation. However, as noted by U.S. EPA (2006), a small amount of hydrogen sulfide (H_2S; <10% of the total gas) is produced. The presence of H_2S is indicative of sulfate reduction and it is known that sulfate-reducing bacteria can be efficient Hg methylators. So, if active sulfate reduction is occurring (typically sulfate concentrations are not limiting in WWTP waters) then there is the potential for Hg methylation to occur within the plant. Other organisms that have been implicated in Hg methylation are iron-reducing bacteria (Kerin et al., 2006). It is not known whether these organisms play an important role in organic matter digestion in WWTP digesters. Clearly, the issue of the potential for methylation of Hg within the plant should be examined further for facilities, especially where there is an indication of relatively high effluent MeHg and/or %MeHg.

Available data for influent and effluent Hg and MeHg at the OM plant showed no strong seasonal trends, although the effluent data was highly variable (range ~2-30% as MeHg). Neither do the data for two facilities in Long Island Sound (LIS) show any seasonal trend in %MeHg in the effluent (Balcom et al., 2004). These data are also quite variable. Other data on %MeHg in effluent from WWTP's show similar variability. Overall, therefore, there do not seem to be strong seasonal differences in terms of %MeHg in the effluent.

Overall, there appear to be some systems where methylation is being enhanced, as demonstrated with the Onondaga Metro facility, and suggested by the other systems with high %MeHg (assuming that the input %MeHg is similar for all systems (1-3%)). Thus, it appears that there are conditions within WWTP plants that may lead to enhanced Hg methylation. Overall, any enhancement of methylation will result in enhanced bioavailability of Hg in the effluent upon discharge into the receiving waters.

3.2.3 Dissolved-Solid Partitioning

The relative bioavailability of solids-bound Hg compared to dissolved Hg is discussed in this section. Estimates of partitioning of Hg between the dissolved and solid phase give an idea of the relative strength of binding of Hg to solid material in WWTP effluents compared to receiving waters. The limited data available from the LIS study show that the K_Ds for total Hg are similar to other waters (Table 3-3), although one plant in Table 3-3 has a higher estimated value. The K_D for the SJ/SC facility was estimated from the solids concentration, the final TSS and the final effluent dissolved Hg concentration ($\log K_D$ of ~6). The relationship between TSS and total Hg by Hsu (2004) was used to estimate a $\log K_D$ of ~5 for the Sacramento Regional facility. These data suggest that the partitioning of Hg to the solid phase in the effluent, while showing some variability, is not substantially different from the range found in receiving waters which shows similar variability across systems. Thus, it is likely that similar processes control partitioning in all waters.

Overall, the data on Hg removal across WWTPs and the relationship to K_D is consistent with the data of Shafer et al. (1998) who showed a strong correlation between these two parameters for 12 metals, with silver (Ag), the metal with the highest K_D in their study, similar to that of Hg, having the highest removal efficiency across the plant. Clearly, if there was some factor that would enhance the partitioning, it would have the overall effect of reducing the filtered Hg loading from the facility and would enhance overall Hg removal. More investigation is required to ascertain if K_D relates to any particular parameter in the effluent. Likely candidates are the organic and sulfur content of the solid material, the dissolved organic content of the effluent, and the colloidal content of the filtered fraction.

Table 3-4. Estimated Partition Coefficients for T-Hg and MeHg in WWTP Effluents.

Facility	Log K_D T-Hg	Log K_D MeHg	Reference
Mattahasset-Cromwell	5.16 ± 0.35	5.18 ± 0.21	1
Hartford	5.10 ± 0.18	4.98 ± 0.21	1
Greenwich	6.7	5.0	1
New Haven	5.1		1
Norwalk	6.7		1
Stamford	6.0		1
San Jose/Santa Clara	~6.0		2
Data from Hsu (2004)	4.9*		3
Receiving Waters			
Connecticut River	5.14 ± 0.42	4.5 – 5.5	1
Surface Fresh Waters	4.0 - 6.0	3.5 – 5.5	5, 7
Surface Estuarine Waters	4.9 – 5.7	3.5 – 5.5	4, 5, 6

Values with error bars represent multiple measurements in the same facility

* Based on the regression slope between total Hg and TSS.

Sources: 1= Balcom et al. (2004); 2 = ESD (2007); 3 = Hsu (2004); 4 = Cossa et al. (1996); 5 = Mason and Benoit (2003); 6 = Mason et al. (1999); 7 = Lawson et al. (2001). For comparison, values for receiving waters are also shown.

Furthermore, the data of Shafer et al. (1998) for Ag is instructive toward what may happen with Hg. In their investigation of a number of WWTPs in Wisconsin, they performed partitioning studies at two plants and found, on average, about 18% of the Ag was in the "dissolved" fraction (<0.1 μm), 23% in the "large colloidal" fraction (0.1 – 0.4 μm) and 60% in the >0.4 μm fraction (the typical particulate fraction). Of the Ag in the "dissolved" fraction, ultrafiltration studies showed that most of this was colloidal (66% in the 10,000 molecular weight (10 kDa) to 0.1 μm fraction). Overall, only about 2% of the total Ag was in the <10 kDa fraction, a very small fraction of the total Ag. Other metals (Cu and Co) were less strongly associated with the colloidal fraction. These metals have much lower K_Ds compared to Hg and Ag.

In the LIS WWTPs, a similar fraction of the total Hg was in the <0.4 μm fraction (40-50%; Balcom et al., 2004), confirming the similarity in the behavior of these two metals, also reflected in the similar overall K_Ds. Thus, one may speculate that a similar fraction of the Hg, about a quarter of the total Hg in the effluent, is in a colloidal phase (0.1 – 0.4 μm), and that the majority of the Hg in the <0.1 μm fraction is likely associated with colloids (10 kDa to 0.1 μm).

There is less information on K_D for MeHg in WWTP effluent, but the information available, mostly from LIS facilities (Balcom et al., 2004), suggests that it is high, on the order of 10^5. This is also consistent with the data of Shafer et al. (1998) in terms of its relative removal, suggesting that MeHg partitions strongly to particles within the plant, is likely strongly removed within the plant (>80%) and is perhaps even more strongly bound to wastewater treatment solids particles compared to particulate matter in natural waters (Table 3-3). While there is little information on differences in K_D for MeHg among different WWTPs, there is likely to be similar variability as found for total Hg. It is likely, however, the same processes that enhance partitioning of inorganic Hg to particles are also responsible for the strong partitioning of MeHg, and thus any process that is found to enhance inorganic Hg removal is likely to enhance MeHg removal as well.

As will be discussed in more detail below, the relative bioavailability of the various fractions of Hg and MeHg should be different. If bound to the solid phase, Hg and MeHg must be released into solution before they can be methylated or bioaccumulated. This is true even for higher trophic level organisms that may ingest solids (algae, sediments, or organisms) as it is thought that only soluble forms are taken up across gut membranes. To a large degree, the factors that enhance the binding of MeHg to the solid phase also result in less solubilization of MeHg from solid material during digestion (Mason (2002) and references therein). Therefore, the strength of binding can provide some assessment of the relative bioavailability of MeHg upon digestion, as well as an indication of the amount of MeHg that will be in solution. The potential bioavailability of Hg and MeHg in solid material upon digestion was discussed in some detail in Chapter 2.0.

Overall, adsorbed-phase Hg and MeHg in the environment will only become bioavailable if there is degradation of particles or a change in the water and/or sediment chemistry that will change the partitioning of Hg to the solid phase, or if the Hg and MeHg is released from the solid phase during digestion by organisms. Thus, compared to other forms, the Hg and MeHg bound to particulate are likely to be the least available in terms of uptake into the base of the food chain (algae and bacteria). Colloidal material is likely to be more easily degraded, but it is still relatively stable, as discussed below, and may indeed coagulate on mixing with the receiving waters and be removed to the sediment as particulates. While it was noted above that the Ag was less strongly bound to colloidal material as compared to particles, it was still strongly bound and likely would be relatively unavailable for bioaccumulation. The same is likely true for Hg. Thus, the most bioavailable fraction, as noted in many other studies, is the dissolved fraction that can be readily accumulated into microorganisms. As noted above, for most effluents this is a relatively small fraction of the total Hg, unless this changes on mixing of the effluent and the receiving waters.

If the notion of equilibrium partitioning is accepted, then the dissolved concentration should not be a function of the concentration of particulate material, only the Hg concentration on a per mass basis. However, there is some evidence in the literature that this is not always the case, and that the measured K_D is a function of the TSS concentration. A well recognized example of this is in EPA's metals translator guidance (U.S. EPA, 1996), where K_D is calculated as a nonlinear function of TSS. This is most likely a result of the fact that colloidal material passes the filter and is therefore registered as "dissolved" and that the amount of colloidal material is related to the amount of TSS. Thus, there could be a higher filtered Hg concentration in effluents that have a higher TSS and a higher total Hg load. However, the LIS data does not support such a conclusion (Figure 3-2b). Therefore, unless clearer evidence is found in other studies, it is not possible to conclude that the filtered Hg concentration is a strong function of the TSS concentration.

Overall, reducing TSS in the final effluent will result in reduced Hg mass in the final effluent. Removal of particles is therefore one of the most effective ways to reduce total Hg and the mass of potentially bioavailable Hg in the receiving waters. A number of studies of Hg concentration and fate across sewage treatment facilities (e.g. the Onondaga Metro and SJ/SC facilities) (Figures 3-4a,b) have shown that the greatest removal of Hg occurs during primary treatment and a much smaller fraction is removed with subsequent processing.

3.2.4. Total Mercury Speciation and Water Quality Parameters

Major Ions: The study of the SJ/SC facility found no association with Hg concentration and that of two major ions – chloride (Cl) and sulfate (SO_4). The Cl concentration appeared to be relatively conservative across the treatment plant and ranged from around 150 to 200 mg/L. However, it is possible that at some locations, and especially in winter, higher Cl concentrations could occur due to the use of salts in raod de-icing and these higher intermittent concentrations could have a short-term impact on Hg speciation within a WWTP plant. The sulfate concentration was also relatively consistent across the plant, being on average between 100 and 120 mg/L. Sulfate values of this magnitude are such that sulfate-reducing bacteria, if present in the plant, would not be sulfate limited. There is little evidence in the total sulfate concentrations across the plant of any substantial depletion that would signify significant conversion of sulfate to sulfide by sulfate reduction. It may be that there are limitations on this process in this WWTP, which might be of potential significance in some facilities.

More information should be gathered on the concentration of sulfate in WWTP effluents. While there may not be an impact of sulfate on methylation within the plant, potentially higher sulfate concentrations upon mixing may lead to an enhancement of methylation in receiving waters. If sulfate levels in the SJ/SC plant are typical, then they would not substantially change

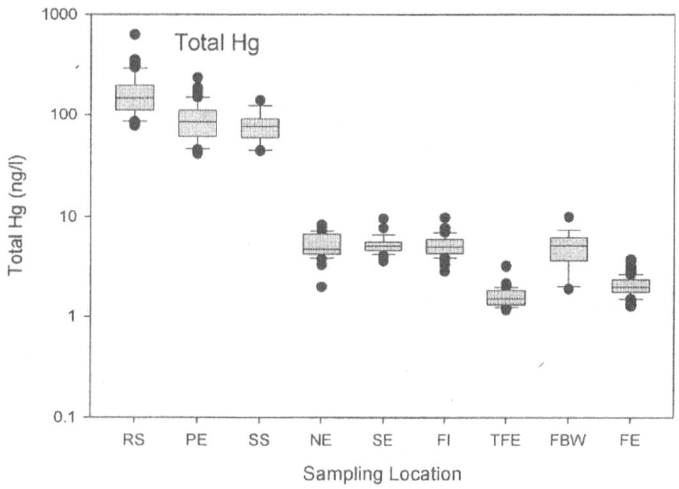

Figure 3-4a. Change in Total Mercury Concentration across the San Jose/Santa Clara Wastewater Treatment Facility. Reprinted with permission from Environmental Services Department (2007).

the concentrations of sulfate in the receiving waters at typical freshwater sulfate concentrations (10-20 mg/L) if mixing ratios are low (1:10). However, this may be a concern in effluent dominated and/or low sulfate receiving waters. Clearly, sulfate is not an issue for discharges into brackish or saline waters.

Chloride concentration could be of significance if it were to alter the dissolved speciation of Hg and MeHg in the effluent or receiving waters, as it has been shown that neutral complexes of Hg and MeHg with Cl are relatively bioavailable. However, given the relatively low Cl concentrations and the typical dissolved organic matter (NOM) concentrations in effluents, estimates of the Hg complexation under these conditions suggest that Cl complexation is not important. For example, in the SJ/SC effluent, where Cl is 5mM (~180 mg/L), given a typical NOM level of 5 mg/L and Hg binding constants to NOM of Benoit et al. (2001), one would estimate that most of the Hg would *not* be in the form of the Cl complexes, but would be bound to NOM. Here again, the Cl concentration in the effluent is not sufficiently different from freshwater values (Cl ~ 10^{-4} M) that it would have a significant effect on the overall composition of the receiving fresh waters in terms of Cl and ionic strength, unless mixing ratios are high. Discharges into brackish and saline waters would not alter the major ion speciation given the much higher Cl concentrations of these waters.

Dissolved Organic Matter and Sulfide: There is little information available on the relationship between Hg in wastewater effluents and NOM concentrations. Here again, however, studies on other metals can be instructive in illustrating what may happen with Hg. Shafer et al. (1998) found a reasonable correlation between filterable Ag and organic matter in the effluents of the plants studied. Given the partitioning results discussed above, this is reasonable if most of the colloidal material in the filtered fraction is NOM. They also found that the K_D for the colloidal fraction was somewhat lower than that for the larger particles, suggesting that the association of Ag to these smaller colloids is weaker, that the larger particles have some Ag that is more strongly bound, potentially in sulfide phases, or Ag is strongly incorporated into the larger particle matrix. Herrin et al. (2001) found a similar association between the degree of Ag complexation and NOM in their study of metal binding ligands in WWTP effluents. Finally, Van Veen et al. (2002) also found an association between Cu complexing capacity of the effluents they studied, the dissolved Cu concentration, and NOM. All these studies suggest therefore a role for organic matter in binding metals in effluent streams. There is insufficient data available for Hg to reach a similar conclusion regarding its fractionation between the filtered and particulate fractions and the role of NOM, but it appears reasonable to conclude it is important. It is likely, as a result, that there may be variable dissolved Hg fractions for systems that vary widely in NOM content. The variation in NOM content across WWTPs needs to be further investigated but it appears to range from around 5 to 15 mg/L (Shafer et al., 1998; Hsu and Sedlak, 2003), although the data of Van Veen et al. (2002) suggest that much higher concentrations are possible (up to 70 mg/L), as found in their study of WWTPs in the United Kingdom.

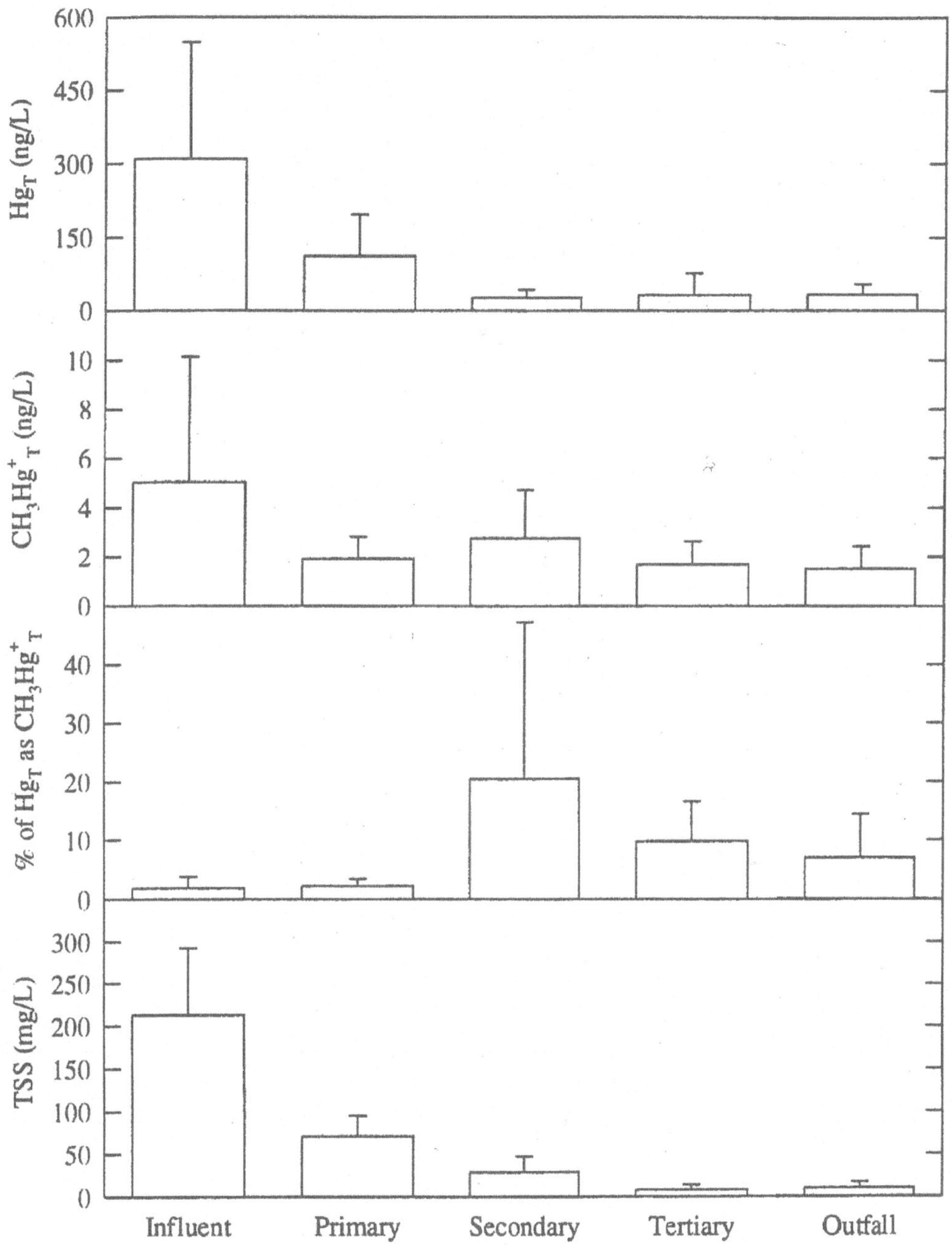

Figure 3-4b. Change in Total Mercury, Methylmercury, Percent Total mercury as Methylmercury, and Total Suspended Solids Concentration Across the Onondaga City Wastewater Treatment Facility. Reprinted with permission from J. McAlear (1996).

The data of Hsu and Sedlak (2003; 2005) also suggest a strong partitioning of Hg to "ligands" present in the filtered fraction of effluent and that these associations are relatively stable to degradation. It is likely that such ligands represent either organic (thiol) binding sites within the organic matrix, metal sulfide entities, or even some combination involving Hg, NOM and sulfide (Miller et al., 2007). If the Hg is being bound to sulfide entities, these could either be small molecular weight, dissolved metal complexes or larger, colloidal metal-sulfide compounds containing multiple metal-sulfide linkages. It is known that Hg binds strongly to sulfide in solution and can exist as numerous dissolved species of the general form $HS_xH_y^{(2-2x+y)}$. If the compounds have more than one Hg atom, such compounds have been termed metal-sulfide "clusters" or "nanoparticles" and have been shown to have the general formula M_xS_y. For clusters, $1<x<10$ (Luther et al., 1999; Rozan et al., 2000) such entities would have a molecular weight of <1 kDa (Luther et al., 1999) and easily pass through most ultrafiltration and other membranes, associated with the "dissolved" fraction. However, if these compounds were of larger molecular weight, they could account for the associations found, for example, by Shafer et al. (1998) for Ag and colloidal material. If such compounds were present it is entirely possible that Hg would interact with the metal clusters or nanoparticles because Hg binds strongly to solid sulfide phases, as shown by its complexation with solid FeS (Miller, 2005). It is not known how the Hg binds to the sulfide, whether through simple adsorption or through precipitation and incorporation into the sulfide matrix, but all such associations would likely result in a highly stable and not easily degraded entity.

Another possibility is that these "compounds" involve an interaction between small Hg-S entities and organic matter. Miller et al. (2007) have shown an interaction between the neutral complexes of Hg-S and NOM in natural waters and laboratory solutions. Neutral Hg-S complexes interacting with NOM in effluent waters would reduce the bioavailability of these complexes, as these have been shown to be bioavailable to bacteria and have been related to the extent of methylation in sulfate reducing bacteria cultures (Benoit et al., 2003). Studies examining the stability and analysis of metal clusters and their bioavailability have found that the stability of these compounds is increased, and their reactivity decreased, in solutions containing NOM (Bowles et al., 2002; Rozan et al., 2000). Thus, it is possible that the stable complexes found by Hsu and Sedlak (2003; 2005) are the result of an interaction of Hg-S clusters and NOM. If the Hg-S complex is within the hydrophobic matrix of NOM, then it is likely to be relatively stable with regard to degradation. Such compounds would fit the results of Hsu and Sedlak (2003) where only a fraction of the bound Hg was titratable with another strong S-containing ligand (glutathione). These authors concluded that there were "nonlabile hydrophobic Hg(II) complex(es)" in the treatment plant effluents at relatively high concentrations. Such complexes were either present in low concentration or absent in natural water samples that were also studied, suggesting that these strong hydrophobic ligands are produced within the WWTP facility. The results of their study showed that all waters had substantial complexing capacity for Hg but that in the WWTP effluents there were a smaller number of very strong binding ligands. These ligands appear to be stable to degradation and thus could be important in the complexation of Hg in the receiving water, both for Hg derived from the WWTP, and that Hg in the receiving water itself.

Studies of Ag binding in effluents reveal that binding constants are on the order of 10^{12} (Smith et al., 2004; Herrin et al., 2001). A value of this magnitude is high for Ag relative to its binding to inorganic ligands and other organic ligands, except those containing thiol groups (Figure 3-5). Similarly, the study of Van Veen et al. (2002) found a high binding constant for Cu. The results of Hsu and Sedlak (2003) indicate a binding strength for Hg of greater than 10^{25}. Lamborg et al. (2004) estimated the binding constant for one effluent to be around 10^{23}, somewhat less than that of Hsu and Sedlak (2003). It is reasonable to conclude that similar ligands are associated with the binding of Hg and Ag in WWTP effluents and a comparison of the binding constants for Hg and Ag are in agreement with the general relationship between the binding constants for these two metals (Figure 3-5). It is worth keeping in mind that in terms of these binding ligands, it may be necessary to consider the other metals, such as Ag and Cu, and perhaps Pb, which can be present in effluents at higher concentrations than Hg as they may compete for these strong binding ligands in the effluent and could reduce Hg binding.

The study of Hsu-Kim and Sedlak (2005) showed that Hg binding ligands formed in WWTPs were stable to oxidation and were similar in behavior to compounds formed between Hg and S-containing organic molecules. However, they were not as labile as Hg complexes formed with NOM, suggesting that the associations are not just those between Hg and the thiol groups of NOM. The strongly Hg-bound complexes, in similar fashion to metal-sulfide complexes and clusters, were destroyed by chlorine disinfection, but were relatively stable in oxygenated solution showing loss of up to 75% of the ligands over a two week period. Thus, over the timescales defined in Chapter 2.0, such compounds would be considered stable. The study of Van Veen et al. (2002) similarly showed that the copper (Cu) complexing capacity of effluents was also only slowly degraded over a 20-day period and that this complexing ability still remained in some effluents, although there was variability in the extent of degradation between effluents. In addition, Van Veen's study showed that mixing of the effluent waters in various dilution ratios did not eliminate the Cu complexing capacity – to the contrary, the amount of complexation after mixing was greater than would be predicted based on conservative mixing, suggesting that there is excess complexation capacity in effluents above that of the metals present. It is likely that the same situation occurs for Hg (Hsu, pers. comm.). In the Van Veen et al. (2002) study, even though 10-70% of the Cu in the different effluents was in the dissolved fraction, their calculations indicate that very little of the filterable Cu was present as inorganic Cu complexes.

All these results suggest that the ligands in question in WWTP effluents are reacting with Hg through S-linkages, and likely involving NOM as well, and are forming complexes or associations including Hg, S and NOM that are relatively stable against degradation or ligand exchange reactions. If this is so, then these compounds will persist in the receiving waters and will not rapidly change their bioavailability. Thus, even though there may be debate over the exact nature of these compounds, it appears that these Hg-containing entities are unreactive and not bioavailable, as discussed further below. Given the information above (e.g Miller et al., 2007; Miller, 2006; Hsu-Kim and Sedlak, 2003; 2005; Lamborg et al., 2004; Benoit et al., 1999a), it is possible to make an estimate of the impact of NOM and dissolved sulfide on the speciation of inorganic Hg in environmental waters. It should be noted that these estimations are based on little data, and the results of studies using particular fractions of organic matter isolated from the environment. Thus these results should not be over-extrapolated. However, they do give some indication of the potential impact of NOM and sulfide on bioavailability. Speciation calculations were performed using the computer program MINEQL, using the constants from literature cited above, and assuming that NOM is solution can be modeled as a spherically charged surface with acid-base characteristics and Hg binding complexation sites (thiols ligands). The results of these calculations are discussed in detail in Chapter 5.0, but the outcome (see Figure 5-4), illustrates that the presence of NOM can have a major impact on the speciation of inorganic Hg in solution.

3.2.5 Relative Bioavailability of Mercury Forms in Wastewater Effluent

In considering the relative bioavailability of Hg in wastewater effluents, it appears necessary to consider three fractions – large particulate (>0.45 μm) containing Hg; colloidal Hg that is within the filtered fraction, and truly dissolved Hg complexes. As particle-bound mercury has been dealt with earlier, attention in this section is focused on truly dissolved Hg and some special considerations for colloidal/particle-phase mercury.

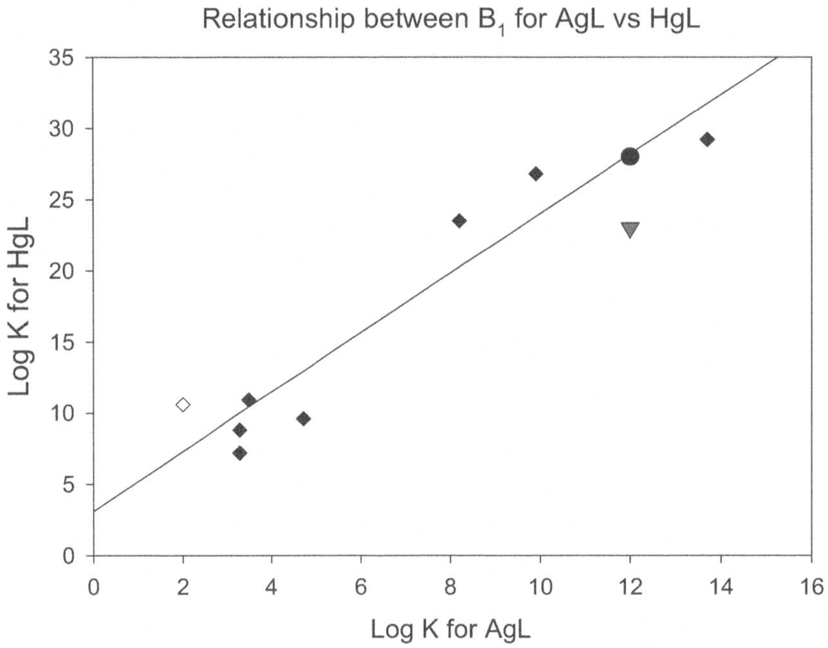

Figure 3-5. Relationship between the Binding Constants of Silver and those of Mercury from the Literature (black diamonds). Also Shown are the points corresponding to the binding constant for Ag binding to the strong effluent ligands (logK ~ 12) and the average value found by Hsu-Kim and Sedlak (2003) (logK ~ 28) (circle) and by Lamborg et al. 2004) (logK ~ 23) (triangle). Data Source: Hsu-Kim and Sedlak (2003), Lamborg et al. (2004).

Dissolved Mercury: Only a very small fraction of the Hg in the effluent stream is likely to be present in a readily dissolved and available form. While this conclusion is based on the foregoing discussions about the role of NOM in Hg complexation, there is little data to support this contention for all WWTPs. However, the study in LIS by Balcom et al. (2004) did measure "reactive Hg" in the effluents; in this case, the amount of Hg reducible by tin chloride in an unfiltered sample acidified immediately prior to analysis. It should also be noted that by this measure, complexes of Hg with some kinetically labile organic molecules will be part of the reactive Hg pool. Lamborg et al. 2003) showed that compounds such as EDTA are reactive according to the reactive Hg metric used. Thus, the fact that the Hg is found to be in the reactive fraction does not prove it is an inorganic complex or the free metal ion but rather is a complex kinetically labile to ligand exchange and reaction. Such a measure is therefore a potential surrogate for the rapidity and degree to which the Hg in the effluent stream may react, re-partition or undergo other chemical changes upon mixing with the receiving waters.

The fraction of the total Hg that was reactive in Balcom et al. (2004)'s study varied widely but was mostly <10% although there were some individual determinations where the fraction was much higher. It is not known what proportion of the Hg associated with particles may be reactive but even for the dissolved fraction, the reactive Hg fraction was mostly less than 10%. These results fit with the speciation data above based on NOM concentration, which suggest that little of the Hg in the effluent pool is in a reactive form (in terms of this chemical reactivity definition). However, measurements of the fraction as reactive Hg in the Connecticut River, the largest river in the region, show a similar overall speciation with about 15% of the total dissolved Hg as reactive Hg. Therefore, it appears that while the Hg in the sewage effluent may be to some degree bound to strong unreactive ligands that may persist in the receiving waters, there is still some fraction of the dissolved Hg that is labile enough, even if these associations are with organic compounds or other ligands, to be determined as reactive depending on the technique used. While these limited results are encouraging concerning the validity of the reactive Hg measurement as a surrogate determination of inorganic Hg bioavailability in WWTP effluents and other soruces, it is recommended that this be further studied before any recommendation is made concerning its use and application in studies of Hg bioavailability. It would be relatively simple to design a suitable experimental study to determine the validity of the reactive Hg measurement in comparison with calculations of bioavailability, as outlined later in Chapter 5.0.

Colloidal and Particulate Mercury: The composition and stability of particulate Hg in terms of bioavailability is likely not to be dramatically affected unless there is a change in the oxidative capacity of the solution upon mixing with the receiving waters, or the particulate material is coagulated and removed to the sediment. Once in the sediment it is possible that changes in redox chemistry and diagenesis may lead to the release of the Hg associated with the particulate to the pore water where it may be methylated or bioaccumulated. The degree to which this may occur will be a function of such parameters as the organic content of the sediment and the degree of binding of Hg to the particles. Clearly, there is the possibility of the Hg attached to particles to be released to solution but this is not likely to occur rapidly. There is currently no data to definitively prove what may occur, although the degree to which the particles will coagulate and flocculate, or disintegrate, will depend on the relative chemistry of the effluent and the receiving waters. The available information suggests that effluents have a high conductivity relative to receiving freshwaters and thus there is the potential for changes in partitioning to occur if there is a change in the ionic strength, and perhaps other variables such as pH, upon mixing.

Interestingly, the study of Shafer et al. (1998) also investigated the fate of Ag upon mixing of the effluent with the receiving waters. In this instance, the effluent had a conductivity of about three times the receiving waters, pH and NOM were similar, but the dilution ratio was about 10:1; therefore the mixing should not have changed the chemistry of the receiving waters in any substantial way. However, they still found non-conservative behavior and a rapid removal of the Ag immediately downstream of the outlet (within 100 m), with losses in a range of 30-60%. High Ag sediment concentrations confirmed the idea of removal of Ag most probably due to coagulation and sinking. Measurements in the water column indicated a rapid increase in particulate Ag and a decrease in filterable Ag, consistent with the other data. Thus, their study appears to provide a case where differences between the effluent and receiving waters, although not apparent in the measurements made, lead to a rapid coagulation and removal of Ag to the sediment downstream of the mixing zone. It is likely in this situation that inorganic Hg would behave similarly. Clearly, more investigation of the possible reasons for coagulation or flocculation, or the opposite, particulate disintegration, need to be examined in order to fully understand the fate of Hg in effluents in receiving water. Data is needed for a variety of different mixing regimes if the complexities of these interactions are to be fully understood.

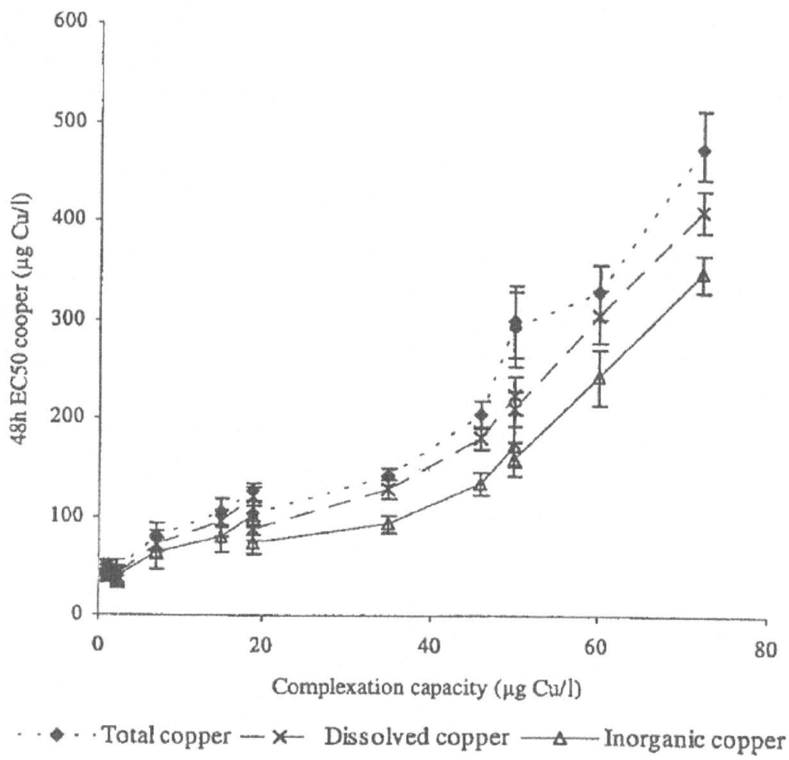

Figure 3-6. Toxicity Data for Experiments with Daphnia and Effluents and their Mixture with Receiving Waters.
Reprinted with permission from Van Veen et al., *Environmental Toxicology & Chemistry*, Allen Press Publishing Services (2002).

While Shafer et al. (1998) found non-conservative behavior, the experiments of Van Veen et al. (2002) found conservative behavior in their laboratory-based mixing experiments. These authors studied the fate of Cu in WWTP effluents upon mixing with receiving waters, the complexing capacity of the effluents for Cu, and that of mixtures of effluent and receiving waters. They also conducted toxicity assays with the effluents themselves and with mixtures of the effluents and receiving waters. As shown in Figure 3-6, the toxicity (EC50) increased with increasing complexing capacity of the waters and mixtures so it was evident that the Cu complex to the ligands was not bioavailable to the test organisms – *Daphnia*. These results confirm the idea that the metals bound up in colloidal fractions, or bound to NOM (as most of the NOM will be in the colloidal fractions), are not as bioavailable as the inorganically bound metal or the free metal ion. This is consistent with theory. This study also showed an "excess" complexing capacity in the effluents. Thus it is possible that the mixing of the effluents and the receiving waters could lead to an overall decrease in the bioavailability of the mixture compared to even that of the receiving waters. This idea that the mixing is non-conservative needs further study.

Any factor that enhances the degree of coagulation of colloidal material and/or particle settling within the mixing zone could lead to the formation of a sediment concentration "hotspot". However, the exact impact of the high Hg deposition will only be important if sediment organic content does not concomitantly increase, as the higher sediment organic content could actually lead to a reduction in the bioavailability of the Hg in the sediment. In many instances, it is likely that coagulation will involve removal of NOM to the particulate phase. The overall result is complex given the dual role of organic matter – it can help trap Hg in sediment and thus increase the total Hg burden of the sediment, but its strong binding capacity results in a high Hg sediment K_D. As noted above, most often systems that have high sediment total Hg are not regions where Hg is efficiently methylated because of the concomitant high organic content of these sediments. Indeed, data for a number of estuarine sediments in the eastern US, which are impacted due to man's inputs, including WWTPs, show an inverse relation between the sediment total Hg and %MeHg, affirming that a high sediment organic content may help limit Hg availability to methylating organisms.

Other Inorganic Forms: There is the potential for elemental Hg (Hg^0) to be present in the effluent, or to be formed upon mixing of the effluent and receiving waters, due to the reduction of ionic Hg (Hg^{II}), mainly by photochemical processes. Conversion of ionic Hg to Hg^0 could lead to a decrease in the Hg load as Hg^0, a dissolved gas, can be lost to the atmosphere if it is present in solution at supersaturated concentrations. This is mostly the case as the concentration of Hg^0 in solution at equilibrium with the atmosphere (typically ~1.5 ng/m^3 Hg^0) is <0.01 ng/L. However, it appears that this form of Hg is not likely to be important as it would be measured as reactive Hg using the measurement approach of Balcom et al. (2004) and so is likely to be <10% in most cases. There could be the potential for enhanced reduction of Hg^{II} upon mixing of the effluent and the receiving waters. Factors that appear to enhance Hg^{II} reduction are low solids concentrations and color (both of which reduce light penetration when elevated) and low Cl, which at higher concentrations would tend to stabilize Hg^{II} in solution. Higher NOM may enhance reduction as Hg-NOM complexes may be "photo-active" although it could also reduce light penetration, so the role of NOM is complex. Overall, however, it is likely that Hg^0 formation and evasion are not important processes for WWTP plant effluents.

Methylmercury: A similar consideration to that given above to inorganic Hg should also be made for MeHg. However, the data is much more limited and more information is needed on partitioning and distribution of MeHg between truly dissolved and colloidal phases before this can be done. For MeHg, it is also possible to contrast the information obtained from study of other metals with similar chemistry, as has been done above.

3.2.6 Impact of Effluent Water Chemistry

While it has been mentioned above, it is worth reiterating the factors besides those directly affecting the speciation of Hg and MeHg in WWTP effluents that could be important in determining bioavailability in receiving waters, particularly in terms of their potential impact on methylation, bioavailability and bioaccumulation. The role of NOM is complex as it is both a strong complexing agent for Hg, thus helping to keep the Hg in the filtered fraction, and potentially more bioavailable, as it may also stimulate the degree of sulfate reduction in the receiving waters by providing degradable organic carbon for these organisms. Thus its role could be two-fold, and its potential effects on bioavailability are difficult to predict.

Sulfate concentration could help stimulate methylation, as discussed above. In addition, nutrient concentration of the effluent may be substantially higher than the receiving waters which may stimulate microbial and algal growth. If this lead to higher organic matter deposition to sediment, it may enhance methylation. As with NOM, the impact of eutrophication is complex as it can result in both positive and negative feedbacks that could influence methylation in different ways. All these factors need further consideration in different scenarios to understand more clearly their potential role.

3.3 Factors Relating to Receiving Waters

3.3.1 Dilution and Biogeochemistry of Receiving Waters

Before discussing the components of the receiving waters that may affect the bioavailability of Hg, it is important to point out that the dilution ratio of effluent to receiving waters is very important. It is not known to what degree it varies across the different facilities and whether there are any trends; e.g. low dilution for small streams. If the dilution ratio is high (i.e. the effluent highly diluted by the receiving waters, >10:1 receiving water: effluent), then the chemistry of the mixture will approximate that of the receiving waters and the effluent chemistry should have little impact on the receiving water chemistry unless the two are radically different. This is unlikely to be the case for variables such as TSS, NOM and solids organic content. However, in the case of ionic strength (conductivity) it appears that effluents have a relatively high conductivity (specific conductance (SC) >1000 μS/cm) as reported in the literature compared to typical freshwaters (lakes <100-500 μS/cm, and rivers somewhat higher), and lowering conductivity via mixing could lead to enhancement of bioavailability. On the contrary, if the dilution ratio is low (>10:1 effluent: receiving water), then the effluent chemistry will likely dominate the chemistry in the receiving water and the factors that were discussed above would be more important in determining bioavailability. If the dilution ratio is in-between these two extremes, it is more difficult to ascertain the overall impact of the receiving water chemistry, and these scenarios would need to be considered individually. Given this discussion, the consideration of the chemistry of the receiving waters in terms of their impact on bioavailability is only pertinent to those systems where their chemistry dominates the final mixture.

The primary chemistry variables in receiving waters that may be important are the same as those important in the effluent. For example, the partitioning of the Hg and MeHg in the receiving waters will depend on parameters such as NOM and organic content of particles, and major ion concentrations. If these constituents are present in relatively high concentration, they could diminish the bioavailability of Hg or MeHg added via the effluent. NOM and organic particle content are unlikely to be lower in concentration compared to the effluent except for cases of discharge into relatively oligotrophic waters. If receiving waters have high sulfate and other conditions that may be conducive to Hg methylation, then an increase in the conversion of inorganic Hg from the effluent into MeHg could result, thus increasing bioavailability. Receiving waters with low oxygen and/or high biological oxygen demand in waters and sediment are environments that could lead to enhanced methylation, and thus higher bioavailability. Clearly, there are some systems that would be much more sensitive than others, as will be elaborated later in this report.

As previously discussed, MeHg concentration of the receiving waters is a function of both the rate of methylation and the rate of demethylation. Any factor that enhances demethylation in the receiving waters will lead to an overall decrease in bioavailability. It is known, for example, that MeHg can be photochemically degraded in surface waters. This process requires the penetration of UV radiation and thus is enhanced in low TSS, low color waters and is less important for high NOM waters or waters with elevated TSS concentrations. Thus, photo-demethylation is not a ubiquitous process and is likely to be important in only a few situations. Biotic demethylation is also thought to occur but the factors influencing this process are not well known, and it is thought to occur primarily in sediments.

3.3.2 Ionic Strength and Coagulation

For higher dilution ratios, higher relative ionic strength could enhance coagulation and sinking of particles in the mixing zone, removing Hg and MeHg from the water column. As noted above, coagulation theory (Stumm and Morgan, 1996) predicts that coagulation is enhanced by high ionic strength. Thus, discharge of effluents into brackish and saline waters could lead to an enhancement of the near-field removal of Hg on particles to the sediment. Large changes in pH and changes in the concentration of some specific ions could also affect the degree of coagulation or the stability of the effluent particles on mixing. For low dilution ratios, relative ionic strength changes are smaller and less an issue, and this parameter should not impact the degree of coagulation to any substantial degree.

Such conclusions should be used with caution, however. In the study of Shafer et al. (1998) of the Jackson, WI plant who found strong removal of Ag in the mixing zone. Field data on the partitioning between phases downstream and measurements of Ag in sediment suggest that there is strong coagulation and settling of particles containing Ag in the mixing zone, within 100 m of the outfall. The ionic strength of effluent in this case (SC 1,500-2,000 μS/cm) is much higher than that of the receiving waters (620-720 μS/cm), but due to the high dilution (>10:1), the overall increase in SC of the mixed waters is small. Similarly, the differences in NOM and pH between effluent and receiving waters were small, so a dramatic change in these parameters upon mixing would not have been predicted. Overall, therefore, it is not clear what is enhancing coagulation in this instance even though it appears to be occurring.

3.3.3 Physical Mixing and Receiving Waterbody Characteristics

It is not possible to make definitive statements about all aspects of the characteristics of the receiving water environment but a few comments on potentially important factors are necessary. While the exact nature of the mixing regime and the receiving waters is important (e.g. depth of discharge, relative discharge rate) it is likely that systems where there is an efficient and turbulent mixing will likely result in more rapid equilibration of effluent and receiving waters, and this will most likely decrease the potential for localized conditions that may affect bioavailability. There may also be seasonal factors that need to be taken into account in some situations, especially those of the more temperate environments where temperature differences may vary significantly. Many locations also have strong seasonality of flow in the receiving waters. In these systems, conditions of low flow may be much more conducive to increased bioavailability, likely through an increase in potential methylation within the receiving waters, compared to the high flow conditions.

Geometry (e.g. size, shape and depth) of the receiving waterbody is likely important in that shallow systems may be more conducive to methylation of effluent-derived Hg, in that the transport of Hg from effluent to the sediment may be relatively rapid. However, deep waters and discharge in deeper waters may also result in conditions conducive to more methylation, especially if these deeper waters are of low oxygen content. Slow flowing systems will enhance the transfer of Hg to sediments. Substrate type (e.g. sediment, rocks, or concrete) will also have an impact of the potential for methylation. While sediments appear to be important sites for Hg methylation, methylation apparently also occurs in rivers and streams with riffle and low sediment content characteristics and the importance of biofilm coatings in Hg methylation is only recently fully appreciated. Thus, it is difficult to be specific about the role of substrate type in controlling Hg methylation. Other factors could also play a role.

Another factor is the potential that a large enough density difference between the effluent waters and receiving waters could hinder mixing. Such situations will result in an increase in the mixing time and would therefore prolong the time period where Hg would persist in conditions dominated by the effluent rather than the receiving waters.

Overall, it should be noted that some ecosystems are highly sensitive to Hg input as they have a high propensity to methylate Hg. While many of the factors that enhance methylation have been outlined above, it is not always clear in all situations why a system is specifically sensitive. Factors that may lead to enhanced methylation potential or high bioaccumulative capacity are low buffering capacity, low pH, low metal binding capacity, oligotrophy, etc. In addition, in many parts of the country there are large differences in temperature on a seasonal basis. It has been shown that water temperature in spring, as impacted for example by lake size is positively correlated with fish Hg levels (Bodaly et al., 1993), suggesting that seasonal effects, and the type of receiving water, could, within a particular location, have a dramatic impact of the extent of Hg methylation in the receiving waters. Discharge into such systems would likely lead to higher accumulation of bioavailable Hg into the food chain. On the other hand, mercury bioaccumulation in more productive systems that tend more towards eutrophy may be lowered due to biodilution[5] effects.

[5] Biodilution refers to the dilution of bioaccumulated mercury in organisms due to the greater availability of biomass into which it may partition.

CHAPTER 4.0

MERCURY SOURCE PROFILES

The purpose of this chapter is to provide profiles for sources of interest with regard to mercury in aquatic environments. Ultimately, the information in this chapter can be used in conjunction with the information in the proceeding chapters to make assessments of the relative bioavailability of mercury from these sources in receiving waters. As pointed out in Chapter 3.0, the characteristics of the source are relatively more important in systems in which the effluent dominates the effluent/receiving water mixture when making a determination of this kind.

There were three important considerations for the development of the source profiles. The first was which sources should be profiled. The second was the profiling of the selected sources with regard to bioavailable and potentially bioavailable Hg. These forms include methylmercury (MeHg), bioavailable inorganic Hg, and potentially bioavailable inorganic Hg species. The third consideration was the profiling of these sources with regard to those conditions which tend to enhance or mitigate the bioavailability of these forms.

The researchers decided it would be most useful to profile the sources which are most likely to be present across a variety of watersheds. It was helpful to think about this issue in terms of point and nonpoint sources of Hg. Nonpoint sources include atmospheric deposition, Hg containing sediments, urban and non-urban runoff. In some watersheds, especially in the western U.S., but also in other geographical settings, runoff from historic gold and Hg mining activities can be a factor. With the possible exception of mining, nonpoint sources would be present in almost every instance where one might compare the bioavailability of wastewater effluent to other sources.

Point sources of interest in some watersheds may include wastewater treatment plants, power plants, chloralkali plants, nuclear weapons plants, refineries, laboratories, and other industrial facilities which use Hg in the manufacturing process. In a report to the Minnesota Pollution Control Agency, Barr Engineering (2001) identified 14 product lines that are potential sources of Hg either because of its desirable physical/chemical properties or because it is a contaminant in the manufacture of the certain products. These product lines include:

- lamps (fluorescent, high pressure sodium, mercury arc, metal halide, neon, UV disinfectant)
- mercury switches
- mercury relays
- timers
- batteries (alkaline, Hg-Zn, Hg-Cd)
- rectifiers
- flywheels

- instruments (manometers, barometers, hydrometers, sphymonometers, thermometers, thermostats)
- mercury-operated flame sensor switches, and
- gas regulators

In addition, Jones et al. (1996) describe a number of manufacturing sources and products that contain mercury including pyrometers, mercury boilers, mirrors, catalysts for the oxidation of organic compounds, gold and silver extraction from ores, rectifiers, cathodes in electrolysis/electroanalysis, and in the generation of chlorine and caustic paper processing, dental amalgams, as a laboratory reagent, lubricants, caulks and coatings, in pharmaceuticals as a slimicide, in dyes, wood preservatives, floor wax, furniture polish, fabric softeners, and chlorine bleach.

With the exception of WWTPs, the researchers chose not to profile point sources. There were two major reasons for this decision. First, although point sources may be important in a particular watershed, they are not as likely to be as ubiquitously present as nonpoint Hg sources. In many US watersheds, and consequently in the TMDLs performed for them, point sources are absent or are not considered. Second, again with the exception of WWTPs, the volumes of point source effluents compared to nonpoint sources are small. Even with larger volume sources, like WWTPs, the mass of Hg they typically contribute in terms of the overall Hg loads to a receiving waterbody tend to be relatively minor. In Hg TMDLs performed by EPA Region 4 for the State of Georgia (e.g. U.S. EPA, 2000), for example, point sources were assumed to contribute less than 1% of the Hg to the current load. These TMDLs further stated that point sources of mercury are quite small with "no discernable impact on water quality"[6]. In the San Francisco Bay Hg TMDL, WWTPs were estimated to contribute only about 2.5% of the Hg mass load to the Bay (RWQCB, 2004). Having made this point, it is also apparent that in some watersheds point sources may be significant, even though there has been considerable progress made toward reducing Hg in wastewater effluents in North America. Certainly in the case of discharge to an effluent dominated stream, the point source would be of primary concern. Guidance on dealing with specific sources in specific watershed settings will be provided in Phase II of this project.

4.1 Sources of Interest

Given the foregoing discussion, it was decided to restrict the profiling to nonpoint sources as this would provide the greatest number of WWTPs with information relevant to making comparisons of relative bioavailability. The sources profiled in this report include the following:

- atmospheric deposition
- mining
- mercury-containing sediments
- urban runoff
- non-urban runoff

---

[6] In the Savannah River watershed, the point sources include, incidentally, a mercury cell chloralkali plant.

- wastewater treatment plants

In the following sections, each of these sources is profiled in terms of MeHg, bioavailable Hg and potentially bioavailable Hg (Section 4.2) as well as the enhancers and mitigators of Hg bioavailability present in these sources.

The research team recognizes that point sources will undoubtedly be important in some settings. It is therefore the research team's intent in Phase II of this project to provide guidance on how to obtain and use information for particular sources in a watershed to make assessments of relative bioavailability. In addition, the profiles developed here are of necessity general. They are intended to be used as default profiles where more site-specific information is not available. For a more detailed assessment of a particular waterbody, it will be helpful to have locally derived information.

Another point worth reiterating is that the focus here is on relative bioavailability and thus far the research dealt with relative concentrations and percentages in this and other sections of this report, rather than in terms of actual load. While this context is essential in order to make comparisons across sources and ecosystems, the fluxes of various sources to a particular receiving waterbody will ultimately be an essential consideration. That is, regardless of the relative bioavailability of a particular source, its importance will ultimately depend on the mass input (load) of that source relative to others. Similarly, the input of these sources relative to the flow or the volume of the receiving water is of ultimate importance. The implications of mass loading, dilution, mixing, and the characteristics of receiving waters relative to these various sources will be considered in the next chapter. Even so, the reader should recognize that the goal of this project is to provide a methodology in which relative bioavailability can be defined and used to enhance understanding of the impact of adding various forms of Hg from a variety of sources to receiving waters rather than presenting detailed models or processes for estimating Hg loadings to, or concentrations in, receiving waters. The researchers foresee the end product of this research providing a means to overlay considerations of relative bioavailability onto existing processes and tools used for these purposes.

4.2 Bioavailable and Potentially Bioavailable Mercury Source Profiles

When thinking of bioavailable Hg, one tends to think of MeHg. However, as pointed out in Chapter 2.0, for the purposes of this report it was necessary to expand this definition to include bioavailable and potentially bioavailable inorganic Hg. The need to consider inorganic Hg arises from the fact that in many ecosystems that have been studied, in situ methylation and production of MeHg is the major source in the receiving water rather than input of MeHg from sources. In terms of the sources profiled below, it was possible in general to develop characteristics of each with regard to MeHg. However, for bioavailable or potentially bioavailable inorganic species, more often than not this was not possible. In a very few instances, "reactive" Hg was determined in studies of a particular source, but this was the extent. Reactive inorganic Hg (R-Hg) was defined in Chapter 2.0 and in general includes species of Hg^{II} that are reducible to Hg^0 by the addition of $SnCl_2$, a mild reductant, prior to analysis. These species are mostly in the dissolved phase but in some instances can also include R-Hg associated with particles, and include inorganic and labile organic complexes of inorganic Hg. As noted in Chapter 2.0, this technique does not measure the MeHg fraction.

Table 4-1 attempts to compare %R-Hg and %MeHg for the various sources reviewed and summarized in this chapter[7]. While it is difficult to place a great deal of certainty in these comparisons, the researchers believe that some logical characterizations have emerged for MeHg. For R-Hg, there is little information and the characterizations are much less certain. Again, however, certain trends and relationships appear evident. With regard to several sources, there is enough evidence to suggest that the R-Hg pool is substantially greater than the MeHg pool.

Table 4-1. Characterization of Reactive Inorganic Mercury (R-Hg) and Methylmercury (MeHg) as a Percent of Total Mercury from Mercury Sources.

Source	% R-Hg	%MeHg
Atmospheric Deposition	15 to 95	1 to 5
Mining Sediments	< 5	< 1
Mining Runoff	< 5	< 1
Mercury-containing Sediments	< 2 [a]	< 1 to 5
Sediment Pore Waters		
Impacted	Not known	>10%
Unimpacted	<20%	>10%
Runoff - urban	5 to 10	1 to 2
Runoff - nonurban	~15 [b]	5 to 15
Wastewater Treatment Plants	5 to 25	1 to 10

[a] Based on the methods employed by Marvin-DiPasquale where tin chloride is added to a sediment slurry and the reactive Hg determined after a predetermined sparging period.

[b] Based on the data of Balcom et al. (2004) cited in Chapter 3.0.

It appears that mining sources have among the lowest %MeHg levels, typically 1% or less. This may be due to a variety of environmental conditions that prevail at most of these sites (and their mine drainage waters and sediments), where either high gradient streams produce well-oxygenated waters, or high sulfide levels inhibit methylation, or due to the fact that a high proportion of the Hg in the solid phase is HgS solids (mercury mines), which are relatively insoluble. Additionally, some fraction of the Hg in sediments is likely to be elemental Hg (Hg^0) (gold mines), which would need to be oxidized before it could be methylated. High concentrations of inorganic Hg may inhibit sulfate reducing bacteria or may induce the activity of MeHg degrading organisms.

---

[7] The literature reviewed for this project was identified primarily through the use of key words searches of computerized databases. Scientific papers and journal articles found in this way were supplemented by the personal knowledge and libraries of the Principal Investigators. The Internal Peer Review Team (IPRT) also identified key sources that had not been otherwise discovered and reviewed. References to support this summary are provided in subsequent sections.

Urban runoff appears to have the next lowest levels of %MeHg, although there a relatively few urban runoff studies that have measured MeHg, and confidence in this range is relatively low. Mercury-containing sediments appear to have the next highest levels of %MeHg, typically ranging from < 1 to 5%. Precipitation surprisingly has characteristics levels of %MeHg of 1 up to 5%. The source of MeHg in precipitation is not well understood but open ocean precipitation appears to have much lower levels of MeHg, suggesting that the source is continental and likely related to particulate scavenging by precipitation. Relative R-Hg concentrations appear to be high for precipitation and this may be due to the relatively low particulate or organic carbon content of rainwater which may tend to make the inorganic fraction more bioavailable. Scavenging of atmospheric reactive gaseous Hg (RGM) which is highly soluble and reactive, may explain elevated percentages of R-Hg in precipitation. Over the ocean, R-Hg is a higher fraction of the total (typically >50%) than found for terrestrial rain. Overall, compared to other sources, inorganic Hg in wet and dry deposition is present in a reactive and highly bioavailable form.

Wastewater treatment plants appear to have the next lowest levels of %MeHg, although they appear to be more variable in this regard than most other sources. As discussed in Chapter 3.0, levels of MeHg in wastewater effluent are undoubtedly impacted greatly by the design and operation of individual plants. In particular, the presence of anaerobic processes in WWTPs (e.g. recycling of sludge digestion supernatant) *may* explain elevated levels in *some* cases, as was suggested in the SCRSD MeHg study. WWTPs have the potential to have a higher relative concentration of R-Hg due to the typically low solids content in the effluent. However, there is evidence, presented in Chapter 2.0, that WWTP effluents may have a higher percentage of strong Hg binding ligands, which may be a combination of organic matter-sulfide complexes and which may reduce the bioavailability of Hg^{II} in effluents. These complexes may also be quite resistant to degradation, even in oxic receiving environments.

Non-urban runoff appears to have the highest %MeHg levels, falling for the most part in a range from 5-15%. This latter result may reflect the presence of bogs and wetlands in many natural watersheds which tend to elevate MeHg relative to other sources. Due to the relatively larger percentage of MeHg in non-urban runoff, one might expect a relatively lower percentage of inorganic R-Hg; however, this speculation is based on little study – the results of Balcom et al. (2004) for the Connecticut River suggest that dissolved R-Hg concentrations were about 15% of the total dissolved Hg.

While T-Hg concentrations are not the primary concern of this project, it is of interest to note that sources with the highest T-Hg appear to be those with the lowest MeHg. McKee et al. (2004) collated total Hg concentrations from 34 watersheds around the world as shown in Figure 4-1. According to the authors, maximum T-Hg concentrations of 1-18 ng/L are characteristic of watersheds with pristine "open space" land use and reservoirs with forested or open space catchments. Concentrations of 8-90 ng/L are typical of mixed land uses comprised of urban, agriculture, and open space and range between 30 and 90 ng/L when urban runoff dominates land use within a watershed. Concentrations in excess of 100 ng/L occur in watersheds where there are specific Hg uses such as heavy industrial sites and historic gold and Hg mining areas. The research team agrees with the author's assessment that runoff as concentrations generally follow the pattern mining sources>urban runoff>non-urban runoff. The characterization of %MeHg would follow an inverse pattern; namely, mining sources show the lowest %MeHg, urban runoff shows an intermediate %MeHg, and non-urban runoff shows the highest. It is also of interest that Kamman et al. (2005) observed the same tendency in rivers, reservoirs and lakes.

In their study, described in greater detail below, the waterbody type exhibiting the highest T-Hg concentrations exhibited the lowest %MeHg, and vice versa.

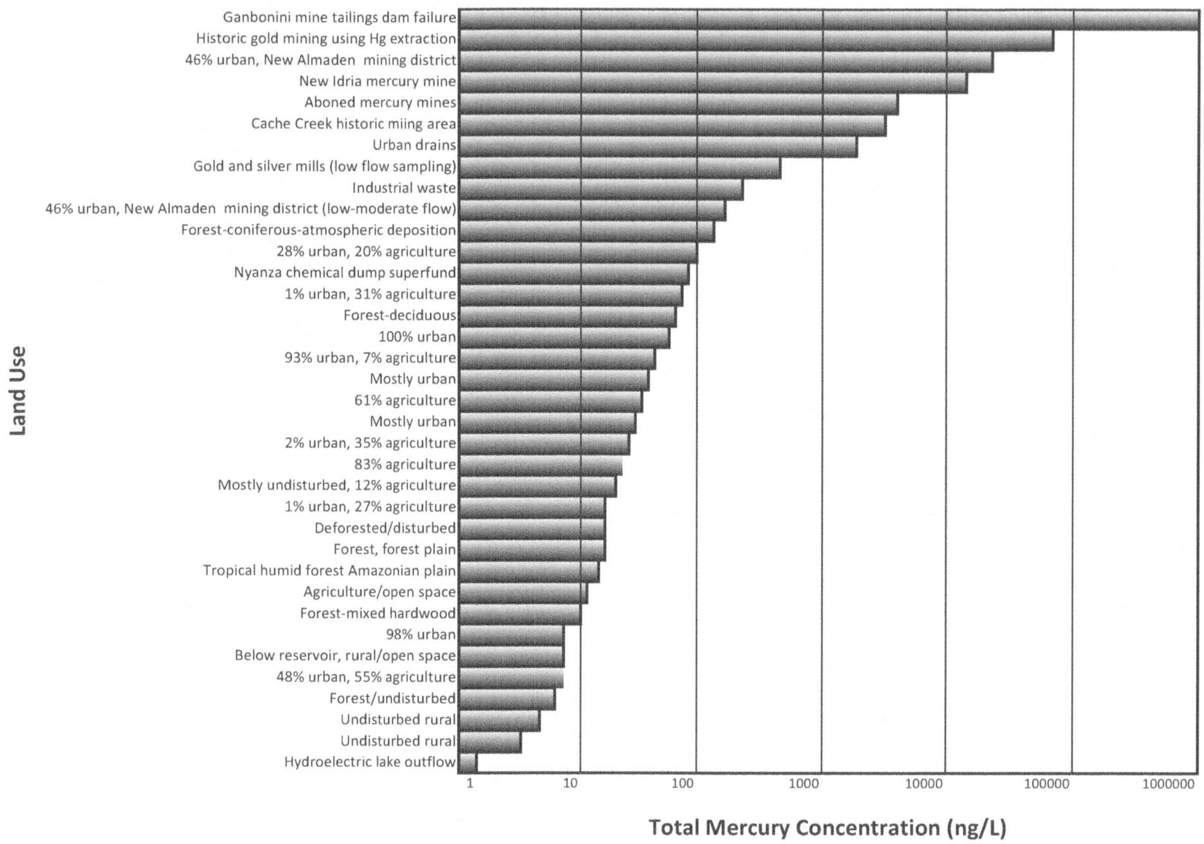

Figure 4-1. Summary of Total Mercury Concentrations in Drainage Waters of Various Land Uses.
Reprinted with permission from McKee et al., San Francisco Estuary Institute (2004).

Another compilation of data (Figure 4-2) highlights the relationship between T-Hg and MeHg in a different way (Benoit et al., 2003). Note that while this log-log relationship is statistically significant it also shows that there must be other factors that influence the extent of net MeHg formation in sediments. For example, at a total Hg concentration of 100 ng/g, it is seen that the MeHg concentration found across ecosystems ranges over at least two orders of magnitude, and that this is true even within a particular ecosystem type (e.g. lakes). As noted above, for most of these ecosystems, it is not the input of MeHg or T-Hg from external sources that accounts for this difference. Rather, the factors controlling *in situ* methylation have as large an effect on the extent of MeHg production as do the differences in total Hg concentration between sites. It is also worthwhile to mention that while a linear relationship is plotted for these data, an exponential fit to the data is also possible. Assuming such a relationship, at low total Hg the slope would be much higher than it is for the more highly contaminated Hg sites. This is discussed further below.

The following section begin with reference to a table in which the information collected for each source is summarized. Following the table, each section contains a more detailed summary for each reference.

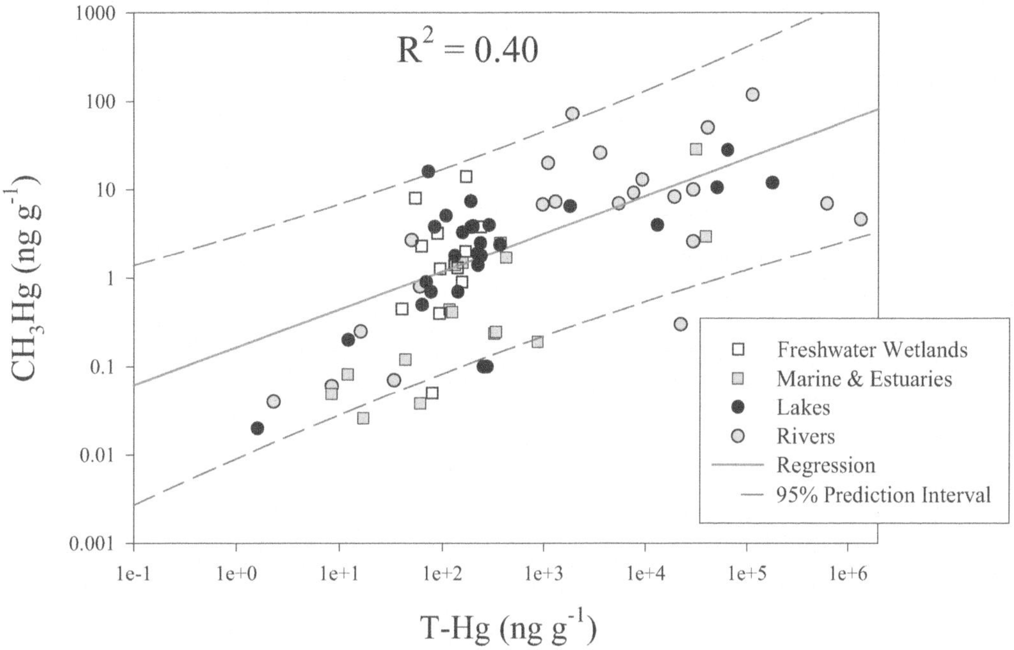

Figure 4-2. The Relationship between Total Mercury and Methylmercury in Sediments for a Variety of Ecosystems from Both Remote and Impacted Locations. Reprinted with permission from Benoit et al., *Biogeochemistry of Environmentally Important Trace Elements,* American Chemical Society (2003).

4.2.1 Atmospheric Deposition

Table 4-2 contains information on total Hg, MeHg and %MeHg in precipitation from various worldwide locations. In general, total Hg in rainfall may range up to 100 ng/L. Methylmercury concentrations range up to 1 ng/L but are typically much lower, with < 1 ng/L appearing to be the norm. As a percentage of total Hg, it appears safe to say that an upper limit of about 5% is appropriate for MeHg in rainfall. In an overview paper, Grigal (2002) reported that MeHg is typically on the order of 1.5% of T-Hg and Downs et al. (2004) report that relative MeHg levels in precipitation average from 1% (somewhat lower in industrial areas) to 5%.

Table 4-2. Representative Concentrations of MeHg and %MeHg in Precipitation Reported in Various Studies in the U.S., Canada, and Western Europe.

Region/Country	T-Hg (ng/L)	MeHg (ng/L)	%MeHg	Reference
Sweden	-	0.37	3.1	Munthe et al. (2005)
Sweden	-	0.34	-	Bishop et al. (1995)
Sweden	7 to 90	0.05 to 0.6	0.94	Lee and Iverfelt (1991)
Baltic Sea	-	ND to 0.32	1 to 4	Holz (1999)
W. North Carolina/U.S.	6 to 25	< 0.025 to 0.16	0.57	Allan and Heyes (1998)
Great Lakes/U.S.	10 to 60	0.01 to 0.85	< 6	Hall et al. (2005)
Chesapeake Bay/U.S.	15		0.5	Mason et al. (2000)
North-central Wisconsin/U.S.	4	0.05	1.2	Bloom and Watras (1989)
Olympic Peninsula, Washington State/U.S.	2 to 5	0.15	3 to 7.5	Bloom and Watras (1989)
Experimental Lakes Area, Northwestern Ontario, Canada	10	0.12	1.2	St. Louis et al. (2001)
Upper Midwest/U.S.	-	0.18	1.5	Glass and Sorenson (1999)
Northern Wisconsin/U.S.	7.9 3.3	.10 .03	1.2 0.9	Lamborg et al. (1995)

A limited number of papers were found in which R-Hg in rainfall was reported. In Mason et al. (1997), R-Hg was found to be about 15% of T-Hg in the Chesapeake Bay region. In data from Wisconsin, Lamborg et al. (1995) reported concentrations of R-Hg in rainfall with means of 3.7 ng/L in one data set and 1.7 ng/L in the other. In these cases, R-Hg was 47% and 51% of T-Hg; however, 44% and 52% of the R-Hg was associated with particulates in rain and snow samples, respectively. As noted above, measurements of Hg and R-Hg in rain over the equatorial Pacific Ocean yielded an average %R-Hg of ~70% and an average total Hg concentration of 2.9 ng/L (Mason et al., 1992). Methylmercury was undetected in these open ocean rain samples (<10 pg/L) and was <1% of T-Hg. In the North Atlantic, Lamborg et al. (1999) found a total average concentration of 3.6 ng/L and an average %R-Hg of 70%. Again, all MeHg concentrations were <10 pg/L (<1% of T-Hg).

In North America, Bloom and Watras (1989) measured MeHg in precipitation from north-central Wisconsin and the Olympic Peninsula of Washington State. Snow samples from north-central Wisconsin contained MeHg levels of about 0.05 ng/L compared to T-Hg levels of about 4 ng/L (1.25%MeHg). A time series of rainfall samples from the Olympic Peninsula showed average MeHg of 0.15 ng/L with T-Hg ranging from 2 to 5 ng/L (3 to 7.5%MeHg based on the average MeHg and range of T-Hg reported). Mason et al. (1997) found that Hg concentrations in wet deposition to a coastal site and an inland site within the Chesapeake Bay watershed ranged from less than 10 pM (2 ng/L) to about 400 pM (80 ng/L), with the bulk of the concentrations being less than 150 pM (30 ng/L). Speciation measurements showed that about 15% was R-Hg and less than 5% on average was MeHg.

Hammerschmidt et al. (2007) summarized the relationship between MeHg and R-Hg in wet deposition at several US locations. In their assessment, the mean molar ratio of MeHg:R-Hg was found to be 0.025 ± 0.006 over a range of MeHg from approximately 0 to 1.2 ng/L. Thus, on average, the concentration of R-Hg was 40 times the concentration of MeHg in rainfall over the continental US. The fraction of R-Hg to total was variable, ranging from 14% to 95% and MeHg in precipitation was reported to be more a function of R-Hg in rainfall than of T-Hg.

St. Louis et al. (2001) reported the volume weighted average of T-Hg in an open area at the Experimental Lakes Area in northwestern Ontario during 1998/1999 to be 10 ng/L. The volume weighted average of MeHg was 0.12 ng/L giving a %MeHg of 1.2%. These levels were reported to be higher than measurements made at the same area from 1992 to 1995 (T-Hg 4 ng/L and MeHg 0.05 ng/L). However, the %MeHg appears to have remained about the same (1.25%). Mason et al. (2000) made measurements of Hg and MeHg in precipitation at the Chesapeake Biological Laboratory (CBL). T-Hg measured over a four year period averaged 14.6 ng/L whereas MeHg over the same period averaged 0.08 ng/L, a ratio of 0.6%. The authors note that the %MeHg value is lower than that reported by Lamborg et al. (1995) for sites in northern Wisconsin (1.2%) and in the Experimental Lakes Area, Ontario, Canada, which averaged 1.3%.

Allan and Heyes (1998) reported dissolved MeHg in precipitation (June-July 1994) at the Coweeta Hydrologic Laboratory in western North Carolina ranging from < 0.025 to 0.16 ng/L. T-Hg ranged from 5.75 to 24.8 ng/L and the geometric mean percentage of MeHg:T-Hg was 0.57 for the events measured. Hall et al. (2005) reported methyl and total Hg levels in precipitation at five sites in the Great Lakes region between May 1997 and December 2003. MeHg concentrations ranged from 0.01 ng/L to 0.85 ng/L. T-Hg concentrations were generally between 10 and 60 ng/L. Percent MeHg was less than 6% at all sites with exceptions of seven events at one site and two events at another where %MeHg ranged between 6 and 18%. Highest MeHg concentrations were generally found in low volume precipitation events. Glass and Sorenson (1999) reported that MeHg in precipitation in the upper Midwest over a six-year period from 1990 to 1995 averaged 0.18 ng/L (1.5% of T-Hg)

Tsai and Hoenicke (2001) reported volume-weighted T-Hg in precipitation for the San Francisco Bay area between August 1999 and November 2000 to be 8 ng/L. However, wet deposition was estimated to account for only 18% of annual deposition with the majority being from dry deposition. Unfortunately in this study no MeHg measurements were made. With regard to dry deposition, most of this will be in the form of reactive gaseous mercury (RGM) which is considered to consist essentially of highly soluble and bioavailable neutral inorganic Hg^{II} complexes such as $HgCl_2$ (Laurier and Mason, 2007).

Worldwide, Munthe et al. (2005) reported that MeHg from November 1991 to April 1994 in an open field site in Sweden averaged 0.37 ng/L and was 3.1% of T-Hg. Bishop et al. (1995) reported MeHg in snow cores in the Svartberget Catchment in northern Sweden to be 0.34 ± 0.17 ng/L. Lee and Iverfelt (1991) found that MeHg in precipitation ranged from <0.05 to 0.6 ng/L at three deposition sites in Sweden. Corresponding T-Hg concentrations ranged from 7 to 90 ng/L. Holz et al. (1999) measured MeHg concentrations precipitation in the Baltic Sea area. MeHg concentrations ranged from below detection up to 0.32 ng/L and made up from 1 to 4% of T-Hg.

4.2.2 Mining

Historic mining activities can be an important source of Hg in some watersheds. Mercury has been mined in many areas of the western Unites States. Rytuba (2002) indicates ten major Hg mineral belts in North America occurring in southwestern Alaska, British Columbia, the Pacific Northwest, California, the Great Basin, Mexico and Arkansas. The North American Hg belts are mainly of the silica-carbonate or hot-spring types. In addition to Hg mining as a source, historic gold mining is a source due to the usage of Hg as an amalgam in the refining of gold. Jones et al. (1996) estimate that approximately 7600 tons of Hg were inadvertently deposited in conjunction with Gold Rush era mining in California. In addition, Hg is found in minor quantities in the ores mined for other heavy and precious metals including both silver and lead. Both of these metals were sought and mined heavily in the western U.S. In the discussions below, the research team separated mines for gold and other metals from Hg mines as the characteristic concentrations of Hg species appear to warrant this distinction. In profiling these sources, they have also provided separate tables for sediments and runoff from historic and abandoned sites.

Table 4-3 summarizes representative concentrations of T-Hg, MeHg and %MeHg in sediments of mine drainages reported in the literature. As context for these profiles, wthe researchers also present three additional studies in the table. Rice (1999) reported sediment concentrations in 541 streambed samples collected across the conterminous U.S. from 1992 to 1996, noting that these are not bulk concentrations but concentrations in the <63 micron size fraction (likely to be the primary fraction where Hg resides). For total Hg, the median concentration was 0.06 μg/g, with the 25th and 75th percentiles being 0.03 and 0.13 μg/g, respectively. The maximum Hg concentration observed was 14.5 μg/g. In an overview paper, Kamman et al. (2005) report T-Hg and MeHg in sediments from 547 river, reservoir and lake sites across northeastern North America. T-Hg concentrations ranged from 0.01 to 3.7 μg/g with mean of 0.19 μg/g. MeHg concentrations ranged from 0.15 to 21 ng/g with a mean of 3.83 ng/g. T-Hg was highest in lakes, intermediate in reservoir, and lowest in river sites. However, relative MeHg concentrations were greatest at river sites (~4.4%) intermediate at reservoir sites (~2.9%) and lowest at lake sites (1.8%). Domagalski (2001) reported sediment concentrations in the Sacramento River, which has historically been subject to both Hg and gold mining activities ranging from 0.01 μg/g to 0.45 μg/g. The highest sediment values came from a sample collected downstream of areas of past Hg mining activities.

In an overview paper, Rytuba (2002) reports that sediments in streams draining Hg mineralized areas contain between 2 to 200 μg/g of T-Hg, with heavy mineral concentrates ranging up to 4,400 μg/g, and calcines from 10 to 1,500 μg/g.

Table 4-3 shows that sediments from gold and other metals mining drainages may range up to about 30 μg/g T-Hg. Although only one study was found that reported MeHg concentrations in sediments of these mining areas (Giddings et al., 2001), MeHg levels were reported to range up to about 25 ng/g. As a percentage of total Hg, MeHg concentrations appear to be rather low; only about 0.2%. In another study from the Philippines, T-Hg was lower but %MeHg concentrations reportedly ranged from 0.1 to 3%. Relatively low MeHg in sediments of mining areas (including both gold and other metals and Hg) appear to be characteristic. This may be attributable to one of several reasons. It is possible that sulfide levels associated with sulfide ores may be high enough to inhibit methylation in these sediments or that elevated Hg levels may induce the activity of Hg detoxifying organisms (with mer genes). It is also possible that in the

Table 4-3. Representative Concentrations of T-Hg, MeHg, and %MeHg in Sediments of Mine Drainages Reported in the Literature.

Region/Country	T-Hg (µg/g)	MeHg (ng/g)	%MeHg	Reference
Conterminous U.S.	<0.02 to 14.5			Rice (1999)
Northeastern North America	0.01 to 3.7	0.15 to 21	4.4% (Rivers) 2.9% (Reservoirs) 1.8% (Lakes)	Kamman et al. (2005)
Sacramento River, CA/U.S.	0.01 to 0.45			Domagalski (2001)
Gold and Other Metals Mines				
Benguet/Philippines	0.0012 to 2.6		0.1 to 3%	Maglambayan et al. (2005)
Northern Rockies/U.S.	0.2 to 5.5			Maret and Skinner (2000)
Silver Creek, UT/U.S.	1 to 27.8	4 to 26	0.16%	Giddings et al. (2001)
Sierra Nevada streams/U.S.	0.05	0.9	1%	Heim et al. (undated)
Streams NC/U.S.	0.2 to 7.4			Callahan et al. (1994)
Streams north GA/U.S.	0.02 to 3.62			Leigh (1997)
Mercury Mines				
Southwestern Alaska/U.S.	up to 5,500		< 1%	Gray et al. (2000)
Almaden Mine, Spain	3 to 2300	0.32 to 82	0.004%	Gray et al. (2004)
Idrija Mine, Slovenia	0.77 to 1347	0.01 to 11	0.0008%	Gray et al. (2004)
Palawan Mine, Philippines	3.7 to 15	0.28 to 3.9	0.03%	Gray et al. (2004)
Idrija River mouth, Slovenia	40	3	0.0075%	Hines et al. (2001)
Marano Lagoon, Italy	4.1 to 6.6			Piani et al. (2005)
Guadalupe River, CA/U.S.	0.41 to 33	1 to 10		Thomas et al. (2002)
Cache Creek, CA/U.S.	0.1 to 7,000			Domagalski et al. (2004)
Humboldt River Basin NV/U.S.	0.17 to 170	0.12 to 0.95	0.0006%	Gray(2003)
Coast Range, CA/U.S.	18	2.1	0.1%	Heim et al. (undated)
Carson River, NV/U.S.		2 to 28	<3%	Chen et al. (1996)

areas around these mines, aquatic stream environments may be well-aerated, thus limiting the (low dissolved oxygen/redox) conditions necessary for sulfate reducing bacteria to methylate Hg.

Sediments in drainages of Hg mines tend to be much higher in T-Hg, with concentrations in the 5,500 to 7,000 µg/g range observed in areas downstream of Hg mines in Alaska and California. Methylmercury concentrations are reported in the U.S. in a range of 0.1 to 30 ng/g. Percent MeHg concentrations are relatively low, with highest levels being typically 1% or less. This also may be due to any of the reasons identified above for sediments of gold and other metal mines.

Waters draining from mining areas were also profiled and are summarized in Table 4-4. This table distinguishes between receiving waters (upper portion of table) and waters that may be more correctly identified as mine drainage water (lower table). Mercury mine drainage waters may contain several thousand ng/L of T-Hg, with MeHg that may range up to 10 ng/L. As with sediments, the percentage of T-Hg, MeHg is typically quite low, with calculated and reported values of typically less than 1%. In receiving waters, T-Hg and MeHg values tend to drop but MeHg appears to be higher as a percentage with values being on the order of several percent. Drainage from gold and other metals mines appear to be lower in both T-Hg and MeHg than drainage from Hg mines, with %MeHg values being similar.

4.2.2.1 Gold and Other Metals Mine Sediments and Drainage Waters

Maret and Skinner (2000) reported T-Hg concentrations in sediment from 16 sites in the Northern Rockies Intermontane Basin Study Area (Clark Fork-Pend Oreille and Spokane Rivers, Washington, Montana, and Idaho), ranged from 0.2 to 5.5 µg/g. They found T-Hg concentrations in river sediments to be positively correlated with production mine density in the contributing watershed (these were gold/lead/copper mines rather than Hg mines).

Giddings et al. (2001) reported MeHg and T-Hg levels in the Silver Creek drainage, near Park City Utah, an area historically used for silver and lead mining. T-Hg in sediments ranged from 1.04 to 27.8 µg/g and MeHg in sediments ranged from 0.004 to 0.026 µg/g. Percent MeHg ranged from 0.06% to 0.4% with a geometric mean of 0.16%. T-Hg in unfiltered water samples ranged from 21.8 to 160 ng/L and MeHg concentrations ranged from 0.06 to 0.39 ng/L. Percent MeHg ranged from 0.19% to 0.56% with a geometric mean of 0.33%.

Gold-bearing ores occur in U.S. locations other than the west, including a band in the Southeastern U.S. that extends along the southern edge of the Appalachian Mountains from central Alabama to Maryland (Leigh 1997). Callahan (1994) measured Hg in stream sediments in the vicinity of four former gold mining sites in North Carolina and found T-Hg values ranging from 200 to 7,400 ng/g. Leigh (1997) measured T-Hg in sediments of streams in the historic Dahlonega gold mining area of North Georgia. Site average values ranged from 0.02 to 3.62 µg/g. MeHg was not measured in either of these studies.

Maglambayan et al. (2005) found 1.18 to 2600 ng/g in sediments at three sites located downstream of small-scale gold mines in the Upper Ambalanga River, Benguet, Philippines where Hg is used for amalgamation in the gold refining process. MeHg concentrations ranged from 0.1 to 3% of T-Hg concentrations.

Table 4-4. Representative Concentrations of T-Hg, MeHg and %MeHg in Runoff Waters Draining Mining Areas as Reported in the Literature.

	T-Hg (ng/L)	MeHg (ng/L)	%MeHg	Reference
Mercury Mine Receiving Waters				
Sacramento River at Freeport, CA/U.S.	3 to 15	0.03 to 0.3	2% [a]	Domagalski (2001)
Coast Fork Willamette River, OR/U.S.	2.2	0.069	3.1%	ODEQ (2004)
Humboldt River, NV/U.S.	2.2 to 6.9	0.02 to 0.21	2%	Gray (2003)
South San Francisco Bay, CA/U.S.	0.76 to 2			Topping et al. (2004)
Idrija River, Slovenia	< 3 to > 30		0.5%	Hines et al. (2001)
Cache Creek, CA/U.S. Large River Sites	up to 275	up to 0.78	0.28% [a]	Domagalski et al. (2004)
Xiaxihe River, China	20 to 310	0.012 to 0.21	0.02% to 0.29%	Horvat et al. (2003)
Gold and Other Metals Mines				
Silver Creek, UT/U.S.	21.8 to 160	0.06 to 0.39	0.33%	Giddings et al. (2001)
Shasta County, CA/U.S.	6.3 (Unfiltered) 1.3 (Filtered)	0.255 (Unfiltered) 0.097 (Filtered)	4.1% (Unfiltered) 7.5% (Filtered)	Ashley et al. (2002)
Mercury Mines				
Carson River, NV/U.S.	7,580	7.2	0.09% [a]	Bonzongo et al. (1996)
Southwestern Alaska/U.S.	500 to 2,500	< 50	< 5%	Gray et al. (2000)
Almaden Mine, Spain	7.6 to 13,000	0.41 to 30	0.23% [a]	Gray et al. (2004)
Palawan Mine, Philippines	170 to 330	< 0.02 to 0.33	0.1% [a]	Gray et al. (2004)
Idrija Mine, Slovenia	6 to 322	0.01 to 0.6	0.19% [a]	Gray et al. (2004)
Cache Creek, CA/U.S. Small Stream Sites	up to 3,000	up to 8	0.27% [a]	Domagalski et al. (2004)
Mine drainages, Humboldt River, NV/U.S.	8.7 to 2000	0.11 to 0.92	0.18%	Gray (2003)
California Coast Range, CA/U.S.	10 to 11,000 (Unfiltered) 0.3 to 3,700 (Filtered)	0.003 to 1.26 (Unfiltered) 0.002 to 0.9 (Filtered)	0.18% (Silica-carbonate mines) 0.026% (Hot-spring mines)	Rytuba (2000)

[a] Calculated from maximum values.

4.2.2.2 Mercury Mine Sediments and Drainage Waters

Macalady et al. (2000) reported MeHg and T-Hg levels in sediments in Clear Lake, CA, a lake historically impacted by Hg mining. T-Hg and MeHg were reported at three lake locations. T-Hg ranged from 2 to 34 µg/g and MeHg concentrations ranged from 0.002 to 0.007 µg/g. Percent MeHg in sediments ranged from 0.02% to 0.1% with a geometric mean of 0.053%. ODEQ (2004) measured T-Hg and MeHg in sediments in the Coast Fork of the Willamette River, which has been historically impacted by Hg mining. T-Hg concentrations ranged from0.026 to 0.902 µg/g. MeHg concentrations ranged from ND at 0.058 ng/g to 1.22 ng/g. Percent MeHg ranged from 0.12 to 0.72% with a geometric mean of 0.33%. Chen et al. (1996) measured T-Hg and MeHg levels in mine-contaminated sediments of the Carson River, NV. Levels of MeHg were reported to range from 2 to 28 ng/g and were less than 3% of T-Hg.

Heim et al. (undated) compared T-Hg, MeHg, and %MeHg in sediments from California Coast Range streams to sediments in Sierra Nevada streams, the difference being that the Coast Range mountain streams are mostly impacted by Hg mines, whereas the Sierra mountain streams are mostly impacted by Hg used as amalgam in gold mining. T-Hg averaged about 18 µg/g in Coast Range sediments and about 0.05 µg/g in Sierra sediments. MeHg was higher in the Hg mine drainages (2.1 ng/g) versus the gold mine drainages (0.9 ng/g). However, %MeHg was higher in the Sierra mountain streams, about 1%, and only about 0.1% in Coast Range sediments. Another fundamental difference in Hg between gold and Hg mines is the form of the Hg source. In gold mining environments, Hg was used as an amalgam and would have originally been in the elemental Hg^0 form; whereas in Hg mines it would tend to have been in mineralized Hg^{II} forms.

As reported in Gray (2004) sediments downstream from the Almaden Mine, Spain, contained 3-2300 µg/g T-Hg and from 0.32 to 82 ng/g MeHg. Sediments downstream of the Idrija Mine, Slovenia, contained from 0.77 to 1347 µg/g T-Hg and from 0.01 to 11 ng/g MeHg. Sediments downstream of the Palawan Mine, Philippines were reported to contain between 3.7 to 15 µg/g T-Hg and between 0.28 to 3.9 ng/g MeHg. Using the ranges reported here and assuming that the highest MeHg values are associated with the highest T-Hg values, the %MeHg for sediments of the Almaden Mine, the Idrija Mine, and the Palawan Mine would be 0.004%, 0.0008%, and <0.03%, respectively.

The researchers found only one study that measured available inorganic Hg in mine drainage sediments. Gray et al. (2000) found T-Hg concentrations of up to 5,500 µg/g in sediments downstream of abandoned Hg mines in Southwestern Alaska. The reported Hg^{II} concentrations to be generally less than 5% of T-Hg and MeHg concentrations were generally less than 1% of T-Hg. The acid extraction performed prior to measurement was analogous to the approach used by others to study "extractable metals" and acid-volatile sulfide (SEM/AVS) in sediments and therefore this value is not comparable to other R-Hg measurements as it solubilizes solid phases and leaches Hg from AVS and other acid-labile solids. The Hg^{II} species appeared to be several times more prevalent than MeHg, pointing out that there may be a potentially bioavailable pool of Hg in such sediments, which could lead to additional methylation if sediments are transported to a more conducive environment.

Topping et al. (2001) reported that concentrations of reactive Hg in sediments at two sites in South San Francisco Bay ranged from 0.38 to 1.96 ng/g. Percent R-Hg relative to total were lower than those reported by Gray above, being in the range of 0.16 to 0.67%, most likely a result of the different measurements made. Measurements of R-Hg by the method of Marvin-

DiPasquale et al. showed that sediment R-Hg was relatively low, typically <2% (Di Pasquale et al., 2007) in the San Francisco Bay and Delta.

Rytuba (2000) reported that concentrations in drainage water are a function of the ore type, with silica-carbonate deposits yielding waters containing T-Hg from 0.005 to 8 μg/L and hot-spring deposits yielding waters containing 0.01 to 11 μg/L. Unfiltered MeHg values ranging from 0.003 to 1.26 ng/L with the highest concentrations associated with silica-carbonate deposits. If mine drainage waters interact with calcines and other mine wastes, T-Hg and MeHg concentrations may increase dramatically, up to two orders of magnitude. Rytuba (2002) further reported that T-Hg in Hg mine drainage water shows extreme variability ranging from less than 1 μg/L to over 200 μg/L.

There is very little information available on R-Hg in either sediments or runoff from mining areas. The nearest measure found of dissolved bioavailable Hg^{II} was "reactive" Hg in some waters. Gray et al. (2000) sampled waters below Hg mines in Alaska. Unfiltered samples showed 500 to 2500 ng/L of T-Hg, whereas dissolved T-Hg were generally less than 50 ng/L. Hg^{II} was typically less than 22 ng/L and MeHg was generally less than 5% of total Hg [8]. As with the sediments reported in this study, Hg^{II} makes up a potentially bioavailable pool of inorganic Hg up around 4 to 5 times greater than the MeHg pool. By comparison, Domagalski (2001) measured reactive Hg in waters of the Sacramento River drainage, which has been subjected to historic gold and Hg mine drainage impacts. $SnCl_2$ reducible (reactive) Hg ranged from <0.1% to 8% of T-Hg.

Bonzongo et al. (1996) found that distribution of T-Hg and MeHg in surface waters along the Carson River was similar to that measured in river bank sediments. Levels of T-Hg up to 7580 ng/L and MeHg up to 7.2 ng/L were in large part due to solids discharged from Hg-contaminated tailings distributed in the river bank sediments.

Gray (2003) reported the results of T-Hg and MeHg sampling in and around abandoned Hg mines in the Humboldt River basin and surrounding areas of Nevada. Sediments in mine drainages ranged from 0.17 to 170 μg/g. MeHg in sediments ranged from 0.12 to 0.95 ng/g. Based on the maximum values reported, %MeHg was 0.0006%. While levels are low, the range is still very large. Sediment samples in the Humboldt River, the receiving water for the mine drainages had lower sediment T-Hg concentrations but higher %MeHg (geometric mean 0.69%). T-Hg in unfiltered surface waters draining two of the mines ranged from 8.7 ng/L to 2,000 ng/L and MeHg in the same samples ranged from 0.11 to 0.92. Percent MeHg ranged from 0.0005% to 10.2% (an even larger range) with a geometric mean of 0.18%. Similar to the trend in sediments, surface water concentrations in the Humboldt River were lower (2.2 to 6.9 ng/L) but %MeHg was higher (geometric mean of 2%).

Hines et al. (2001) reported T-Hg concentrations in water samples from the Idrija River ranged from < 3 to > 300 ng/L with MeHg accounting for about 0.5%. In the estuary, relative MeHg concentrations were about 1.5% of T-Hg. Sediments in the River mouth contained 40 μg/g T-Hg with MeHg concentrations of 3 μg/g (7.5%).

Piani et al. (2005) found levels of 4.1 to 6.6 μg/g T-Hg in sediments of the Marano Lagoon (Italy). Up to 98% of the total Hg was reported to be in an organic form. Upstream sources include a chloralkali plant and the Idrija mine (Slovenia).

---

[8] Similar caveats apply to measurements of R-Hg in waters to those made in sediments, per the previous footnote.

Thomas et al. (2002) reported that sediments in the Guadalupe River, CA range in concentration from 0.41 to 33 μg/g T-Hg (dry weight) and from 0.001 to 0.01 μg/g MeHg (dry weight). This river drains the historic New Almaden Hg mining district. More elevated MeHg concentrations (0.008 to 0.012 μg/g dry weight) were found in sediments deposited immediately upstream of hydraulic structures (e.g. diversion dams and weirs) in the river.

Domagalski (2001) found that the median T-Hg in unfiltered water samples of the Sacramento River drainage basin ranged from <1 ng/L to about 30 ng/L over an annual cycle. T-Hg concentrations were correlated with TSS. $SnCl_2$ reducible (reactive) Hg ranged from <0.1% to 8% of T-Hg. Mean MeHg concentrations measured in unfiltered water samples ranged from 0.1 to 0.2 ng/L. Domagalski et al. (2004) measured T-Hg and MeHg in sediments in the Cache Creek watershed, CA, an area which has a history of Hg mining. T-Hg in streambed sediments ranged from 0.1 μg/g to 7,000 μg/g.

Topping et al. (2004) reported T-Hg sediment concentrations at two sampling stations in the South San Francisco Bay ranging from 233 ng/g to 293 ng/g. Loss on ignition (LOI) ranged from 4.3 to 6.1% and acid volatile sulfides (AVS) ranged from 1.9 to 33.4 μmol/g. Mean water column T-Hg concentrations ranged from 1.1 to 2 ng/L in May and 0.71 to 1.1 ng/L in November.

Ganguli et al. (2000) reported on the speciation of Hg in waters draining from the New Idria mine area in the Coast Range of California. Although unfiltered T-Hg concentrations in the mine drainage water were comparable to those measured above the mine, unfiltered concentrations of T-Hg 1.2 km downstream of the mine were substantially higher, ranging from 2,900 to 12,400 ng/L. Unfiltered MeHg concentrations ranged from 1.1 to 1.7 ng/L downstream of the mine. The geometric mean of the %MeHg in samples from waters below the mine was 0.04%. The percentage of dissolved to total T-Hg varied widely downstream of the mine (0.76% to 10.8%) but had a geometric mean value of 3.2%. Thus the pool of dissolved inorganic Hg was nearly 100 times greater than the total MeHg pool in these waters.

Some information was obtained on the dissolved versus total Hg in runoff water samples and this gives some indication of the potential quantity of Hg^{II}. However, as pointed out in Chapter 3.0, some of the Hg classified as "dissolved", by virtue of the fact that it passes through a 0.4 um filter, may actually be complexed by organic and inorganic ligands, and may not be truly bioavailable.

Data collected by the Oregon DEQ in support of the Willamette River Hg TMDL provide total and dissolved T-Hg and MeHg concentrations of Hg in water downstream of historic Hg mines (ODEQ 2004). Samples were taken quarterly from nine stations over the course of a year resulting in 36 observations. The geometric mean of the unfiltered T-Hg was calculated to be 2.2 ng/L and the geometric mean of the unfiltered MeHg data was 0.069 ng/L, yielding a %MeHg of 3.1%. Dissolved T-Hg and MeHg were also measured. The geometric mean of the dissolved T-Hg was 1.3 ng/L, giving a % dissolved/ total T-Hg of 59%. This study points out that in waters with low levels of particulate matter (TSS concentrations in the Coast Fork were all < 10 mg/L), the dissolved fraction of inorganic T-Hg can be quite large.

Domagalski et al. (2004) measured both total and dissolved T-Hg and MeHg in water samples and sediments in the Cache Creek watershed, CA, an area which has a history of Hg mining. Maximum T-Hg in unfiltered water samples was approximately 275 ng/L at large river sites and 3,000 ng/L at small stream sites. Maximum T-Hg in filtered samples was about 45 ng/L at large river sites and about 400 ng/L at small stream sites. The median ratio of dissolved to total T-Hg over all sites ranged from about 10% to 45%. Maximum MeHg in unfiltered water samples was approximately 0.78 ng/L and about 8 ng/L at small stream sites. Maximum MeHg in filtered samples was about 0.3 ng/L at large river sites and about 7 ng/L at small stream sites. The median ratio of dissolved to total MeHg over all sites ranged from about 29% to 78%. The upper range of the percentage of dissolved to total inorganic Hg compares well to the geometric mean ratio of 59% from the Willamette River. The data also indicate that the percent dissolved/total MeHg can be even higher. This is plausible since it is generally observed that partition coefficients for MeHg are lower than those for T-Hg.

Horvat et al. (2003) measured T-Hg, MeHg, and %MeHg in unfiltered and filtered water samples from the Xiaxihe River in Chian, which has historically been impacted by Hg mining. T-Hg in three unfiltered water samples ranged from 20 ng/L to 310 ng/L while MeHg ranged from 0.012 to 0.210 ng/L. Percent MeHg ranged from 0.02% to 0.295%. Dissolved T-Hg ranged from 3.9% to 23% of unfiltered T-Hg, while dissolved MeHg ranged from 54% to 68%, reflecting the less strong adsorption of MeHg to particles.

Rytuba (2000) reported both filtered and unfiltered T-Hg and MeHg in waters of a number of silica-carbonate and hot-spring type Hg mines in the California Coast Range. T-Hg levels were highest in drainage waters of the hot-spring type mines (0.17 μg/L) and lower in silica-carbonate type mines (0.06 μg/L) and dissolved to total ratios were 19.4% in the silica-carbonate mines and 6.1% in the hot-spring type mines[9]. On the other hand, MeHg was highest in drainage waters of the silica-carbonate mines (0.11 ng/L) and lower in the drainage waters of the hot-spring type mines (0.045 ng/L). Thus unfiltered MeHg to unfiltered T-Hg ratios were higher in the silica-carbonate mine waters (0.18%) than the hot-spring drainage waters (0.03%), both being low compared to natural waters. As a percentage of total (unfiltered) MeHg, dissolved Hg averaged 62.6% in the drainages waters of silica-carbonate mines and 28% from hot-spring mines.

As reported in Gray et al. (2004), unfiltered water samples collected downstream of the Almaden Mine, Spain contained between 7.6 and 13,000 ng/L T-Hg and between 0.41 and 30 ng/L MeHg. T-Hg in a pit lake ranged from 2,200 and 2,800 ng/L and MeHg ranged from 0.040 to 0.065 ng/L. Waters draining the Idrija Mine in Slovenia exhibited T-Hg concentrations between 6 and 322 ng/L and MeHg between 0.01 and 0.06 ng/L. Waters draining the Palawan Mine in the Philippines had T-HG concentrations ranging from 170 to 330 ng/L with MeHg concentrations in the range of <0.02 to 0.33. A pit lake had T-Hg concentrations ranging from 120 to 940 ng/L and MeHg in the range of 1.7 to 3.1. Again, using the maximum value reported for both MeHg and T-Hg, the corresponding MeHg percentages would be 0.23% and 0.0023% for the Almaden Mine drainage waters and pit lake, 0.19% for the Idrija mine drainage waters, and 0.1% and 0.003% for the Palawan Mine drainage waters and pit lake, respectively.

---

[9] All concentrations and percentages reported are geometric means calculated by the authors from data in Rytuba (2000).

4.2.3 Mercury-Containing Sediments

Mercury containing sediments potentially represent a substantial source of Hg to aquatic ecosystems. Topping et al. (2004) calculated based on Hg fluxes from the benthos to the water column in South San Francisco Bay, that Hg in sediments represented a significant source.

Table 4-5 shows representative concentrations of T-Hg and MeHg found in receiving water sediments in rivers, lakes and estuaries worldwide. This table differs from Table 4-2 in that it focuses on receiving waters versus waters that in closer proximity to mining sources. Crustal abundance for mercury is approximately 100 ng/g (Lindsay 1979) and may be considered a background value for T-Hg in sediments. Concentrations of T-Hg are reported up to 140 µg/g in Poplar Creek, a stream impacted by releases from the Oak Ridge, TN weapons plant. Outside of this, values are generally below 35 µg/g, considerably so in most instances. Maximum MeHg levels reported were generally below 35 ng/g. For the most part, the %MeHg relative to total is less than 1 to 5%. In an overview paper, Gilmour and Henry (1991) reported that %MeHg ranged up to 0.21%, 2.6%, 0.81 % and 1.02% in four studies of marine/estuarine sediments. Sediments in 2 freshwater lakes reportedly contained up to 5.3, 12.3 and 16.3%MeHg, while sediments from four additional lakes contained less than 1.5%MeHg.

A recent compilation of data for estuarine systems (Figure 4-3; Mason (unpublished data)) shows more clearly the relationship between total Hg and MeHg for estuarine sediments. These data are comparable to those shown in Figure 4-2 and again show a complex relationship between Hg in sediments and MeHg. One potential factor driving the relationship is the strong relationship between Hg content of estuarine sediments and sediment organic matter content. Many of these high Hg sediments are high in organic matter, are in eutrophic systems and the sediments are sulfidic. All these factors reduce the bioavailability of Hg for methylation, as discussed in earlier chapters, and thus reduce the %MeHg in the more impacted ecosystems.

With regard to sediments, Campbell et al. (1998) reported concentrations of T-Hg and MeHg in Poplar Creek, TN, a creek heavily contaminated by Hg releases from the production of thermonuclear fusion weapons during the mid-1950s and early-1960s. T-Hg concentrations were highest at the mouth of the creek and ranged from 0.63 to 140 µg/g. MeHg concentrations ranged from 1 to 12 ng/g. EPA (2001) reported levels of T-Hg and MeHg in sediments from six South Georgia blackwater rivers. Concentrations of T-Hg ranged from <1.4 to 27.7 ng/g. MeHg concentrations ranged from 0.004 to 0.323 ng/g. %MeHg ranged from 0.05 to 0.71% with a geometric mean of 0.55%.

Colman et al. (1999) reported levels of MeHg and T-Hg in sediments of reservoirs and wetlands of the Sudbury River in East-central Massachusetts. Two reservoirs were reported to have T-Hg concentrations of 0.3 to 0.4 µg/g (Whitehall) and 6 µg/g (Reservoir 2) in surficial sediments. Despite the differences in T-Hg, both sediments had similar MeHg levels of about 15 ng/g. MeHg:T-Hg ratios ranged from 4 to 5% in the surficial sediments of Whitehall Reservoir. In Reservoir 2, the ratios were lower (0.1 to 0.2%). Hop Brook wetland contained 0.24 µg/g in surficial sediments with MeHg/T-Hg ratios of 0.3 to 2%. Kannan et al. (2004) reported levels of T-Hg and MeHg in sediments from estuarine Florida waters. T-Hg levels ranged from 0.001 to 0.219 µg/g and MeHg levels were on average 0.77% of T-Hg.

Heim et al. (undated) reported T-Hg, MeHg, and %MeHg in sediments of valley streams of the Coast and Sierra Nevada Ranges in California. Characteristic concentrations of T-Hg were about 0.1 µg/g in the Coast Range valley streams and about 0.03 in the Sierra Nevada valley streams. MeHg concentrations were not dissimilar, about 0.4 ng/g in the Coast range valley streams and 0.6 ng/g in the Sierra Nevada valley streams. Percent MeHg was higher in the Sierra Nevada valley streams, about 2.5% versus 0.3% in the Coast Range valley streams.

Table 4-5. Representative Concentrations of T-Hg, MeHg, and %MeHg in Hg Containing Sediments of Rivers, Lakes, and Estuaries Reported in the Literature [a].

Region/Country	Degree of Historic Impact	T-Hg (μg/g)	MeHg (ng/g)	%MeHg	Reference
Clear Lake, CA/U.S.	Severe	2 to 34	2 to 7	0.053%	Macalady (2000)
Reservoirs, East-central MA/U.S.	Moderate	0.3 to 6	~15	0.1 to 0.2% 4 to 5%	Colman et al. (1999)
Six Blackwater Rivers, GA/U.S.	Slight	<0.0014 to 0.0277	0.004 to 0.323	0.55%	EPA (2001)
Coast Fork, Willamette River, OR/U.S.	Moderate	0.026 to 0.902	<0.058 to 1.22	0.33%	ODEQ (2004)
Poplar Creek, TN/U.S.	Severe	0.63 to 140	1 to 12		Campbell et al. (1998)
Carson River, NV/U.S.	Severe		2 to 28	<3%	Chen et al. (1996)
Coast and Sierra Nevada valley streams, CA/U.S.	Severe	0.1, 0.03	0.4, 0.6	0.3%, 1%	Heim et al. (undated)
San Francisco Bay, CA/U.S.	Moderate	0.1 to 0.35	< 1 to 3	0.5 to 1%	Heim et al. (2007)
San Francisco Bay, CA/U.S.	Moderate	0.22	0.43	<0.5%	Conaway et al. (2003)
Tagus Estuary, Portugal	Severe		0.3 to 34.6	0.02 to 0.4%	Canario et al. (2005)
Seine River Estuary, France	Slight	0.3 to 1	0.1 to 6	0.1 to 2%	Mikac et al. (1999)
Estuarine Waters, Florida/U.S.	Slight	0.001 to 0.219		<0.77%	Kannan et al. (2004)
Scheldt River Estuary, Netherlands	Moderate	0.7 to 1.4	3.2 to 24	1 to 1.5%	Leermakers et al. (2001)
Mediterranean Coastal Areas	Moderate	0.04 to 12			Stathopoulou et al. (2001)
Delta Region, CA	Moderate	0.1 to 0.2	0.4 to 15	0.7 to 1.4%	Di Pasquale et al. (2007)

[a] Entries in the table represent both impacted and unimpacted sites.

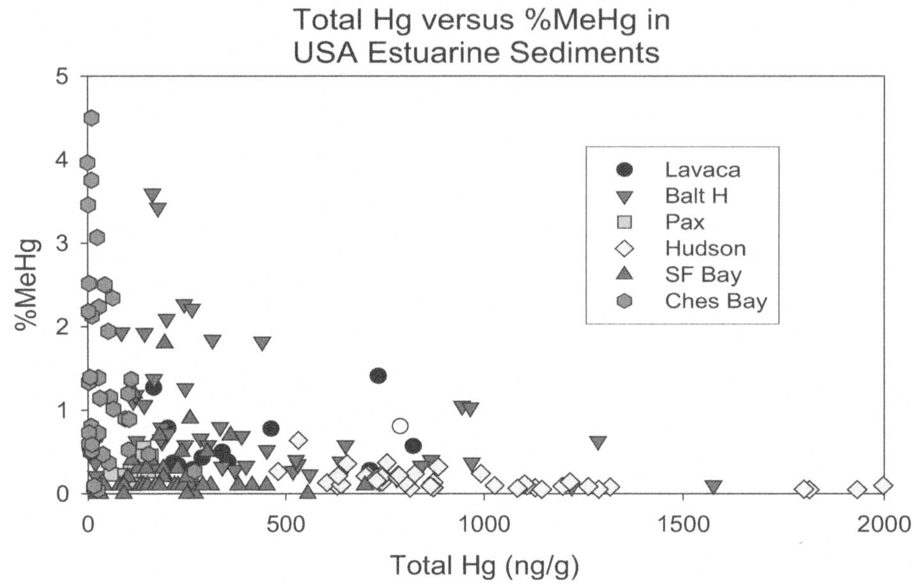

Figure 4-3. Relationship between Total Mercury and Methylmercury for a Number of Estuarine Ecosystems across the U.S. Data taken from Bloom et al. (1999); Mason and Lawrence (1999); Benoit et al. (1998); Heyes et al. (2004); Conaway et al. (2003); Mason et al. (1999) and Heyes et al. (2006).

Heim et al. (2007) found T-Hg concentrations in San Francisco Bay to be typically around 0.3 µg/g (San Pablo Bay and South Bay locations). T-Hg concentrations in the Oakland Inner Harbor were higher, about 1 µg/g with one sample being ten times higher (9.4 µg/g). MeHg concentrations were less than 1 ng/g in San Pablo Bay. The highest MeHg concentration found in the south Bay was 15.9 ng/g. Percent MeHg concentrations in San Pablo Bay and the South Bay ranged primarily between 0.5 and 1%, although two locations had values of 3% and 8%. Conaway et al. (2003) reported an average T-Hg concentration of 0.22 µg/g in San Francisco Bay sediments. MeHg concentrations averaged 0.43 ng/g and the %MeHg was reported to be generally less than 0.5% with a few South Bay locations exhibited relative MeHg concentrations of > 1%.

Worldwide, Canario et al. (2005) reported MeHg concentrations in the contaminated Tagus Estuary, Portugal. MeHg concentrations in surface sediments ranged from less than 1.4 to 160 pM/g (0.3 to 34.6 ng/g). %MeHg to T-Hg was reported to range from 0.02 to 0.4%. Mikac et al. (1999) found T-Hg levels in the Seine River estuary in sediments to range from 0.3 to 1 µg/g. Methylmercury concentrations of up to 6 ng/g were measured and reported to be up to 2% of T-Hg. Leermakers et al. (2001) report sediments of the Scheldt River Estuary to contain between 3.5 to 7 nM/g (0.7 to 1.4 µg/g) T-Hg and 25 to 100 pM/g (5.4 to 21.6 ng/g) MeHg. Relative MeHg was reported to range between 1 and 1.5%. Stathopoulou et al. (2001) reported T-Hg in sediments of coastal Mediterranean areas to range between 0.03 and 12 µg/g.

No studies were found reporting levels of reactive Hg in receiving water sediments. However, several studies reported levels of reactive and MeHg in sediment pore waters. Since Hg in pore water is not sorbed to the solid matrix, it represents a conservative estimate of potentially bioavailable Hg [10]. Nonetheless it represents a departure point for estimating the bioavailable fraction and that fraction that may diffuse into overlying surface waters. Hines et al. (2001) found that T-Hg in pore waters of the Idrija River contained from 1 to 8 ng/L with MeHg accounting for about 30%, a much higher MeHg percentage than found in the same study in sediments. Using the values reported in Tables 4-3 (T-Hg in sediments) and T-Hg in pore waters reported above, the log ratio of pore water T-Hg to sediment T-Hg (the log partition coefficient or $logK_D$) ranges from 6.9 to 7.6. These $logK_D$ values are much higher than found in other freshwater systems (typical sediment $logK_D$s of 4.0 to 5.5 (Mason and Benoit, 2003) and must indicate that much of the solid phase Hg is strongly associated with the sediment matrix – which would occur if the Hg was present in HgS solids – and not available to partition into solution. This may be the case for other sediments associated with mercury mines.

Gagnon et al. (1996) found that MeHg was present at concentrations reaching 10 ng/L in pore waters extracted the anoxic layer of marine sediments but was below detection in pore waters recovered from surficial oxic layers. Dissolved MeHg in the anoxic layers accounted for up to 30% of the dissolved T-Hg but was less than 1% of the MeHg adsorbed to particles. These results are remarkably consistent with those of Hines et al. (2001).

Mikac et al. (1999) reported "very high levels" of T-Hg in sediment pore waters of the Seine River estuary, ranging from 35 to 370 ng/L. Reactive inorganic Hg was also measured in a range of 4 to 100 ng/L. The percentage of labile Hg^{II} complexes available for $SnCl_2$ reduction was reported to vary between 2 and 70%, on average constituting between 15-40% of dissolved T-Hg. Reactive Hg in pore water was compared with other studies. Comparative values were 500 and 800 ng/L for marine and lake sediments, respectively, and 12, 10 and 15 ng/L for coastal, estuarine, freshwater, and sediments, respectively. MeHg in pore waters was generally below the detection level of 2 ng/L. These data suggest that the labile inorganic pool of R-Hg is on the order of 2 to 50 times greater than the MeHg pool in pore water in these sediments. However, this is inconsistent with the other porewater data where MeHg can be a large fraction of the total Hg in porewater. Bulk sediment concentrations ranged between 300 and 1,000 ng/g. MeHg in sediment ranged from 0.1 to 6 ng/g (between 0.1 and 2% of T-Hg) and was reported to be in a typical range for uncontaminated or slightly contaminated estuarine sediments. Using the maximum values for T-Hg and MeHg in sediments and pore water yields $logK_D$ values of 3.43 and 3.48 for T-Hg and MeHg, respectively. These are within the range reported by Colman et al. below.

Colman et al. (1999) reported pore water T-Hg in the sediments of a reservoir in East-central Massachusetts ranged from 23 to 3,400 ng/L and from 1.2 to 420 ng/L in a wetland. MeHg concentrations in pore waters ranged from 1 to 2 ng/L in the reservoir and 4 to 7 ng/L in the wetland surficial sediments. As a percentage, MeHg:T-Hg in pore water ranged from about 14 to 45%, much higher than the percentages reported by Mikac above. Pore water concentrations of T-Hg and MeHg were positively correlated with T-Hg and MeHg in the bulk sediments. Calculated $logK_D$ for T-Hg in pore water given sediment T-Hg concentration in the reservoir would range from 3.2 to 5.4. In the wetland, Calculated $logK_D$ for T-Hg in pore water given sediment T-Hg concentration would range from 2.7 to 5.3.

---

[10] In most cases, pore water Hg is measured on filtered samples so the values represent the dissolved Hg fraction. The bioavailable fraction would be a subset of this if Hg adsorbed to suspended particulates and other complexed forms are accounted for.

Gill et al. (1999) found %MeHg in pore waters of sediments from Lavaca Bay, TX to range between 3% and 80% with the geometric mean being 20%.

4.2.4 Urban Runoff

Urban runoff represents an important and ubiquitous source of Hg loads in many watersheds. Although there is no strict definition for urban runoff, it is generally thought of as runoff originating from urban areas whose land uses consist of residential, commercial, industrial, highways and urban open space.

The research team found only a few recent studies dealing with Hg levels in urban runoff [11]. Of these, two measured Hg in sediments of urban drainages and three measured Hg in water and/or suspended sediments. Of the two key studies that measured Hg in urban runoff one was from the Sacramento River watershed (CEPA, 2005) and the other was from the Washington, DC area (Mason and Sullivan 1998)[12].

Table 4-6 summarizes the key results from these two studies. Although T-Hg in runoff appears to be similar, MeHg values did not appear to vary between low and high flows in the CA study as they did in the DC study. Thus, in the CA study, %MeHg appears to drop at higher flows, while in the DC study it appears to be more consistent between low and high flows.

Table 4-6. Representative Values of T-Hg, MeHg, and %MeHg in Urban Runoff Reported in the Literature.

	T-Hg (ng/L)	% Reactive Hg (ng/L)	MeHg (ng/L)	%MeHg	Reference
Sacramento-San Joaquin Delta, CA/USA	5 (Dry Weather) 25 (Wet Weather)	-	0.2 (Dry Weather) 0.2 (Wet Weather)	4% (Dry Weather) 0.8% (Wet Weather)	CEPA (2005)
Anacostia River, Washington, D.C./USA	6.5 (Base Flow) 35.2 (High Flow)	11.1% (Base Flow) 4.8% (High Flow)	0.095 (Base Flow) 0.58 (High Flow)	1.4% (Base Flow) 1.6% (High Flow)	Mason and Sullivan (1998)

In the Sacramento-San Joaquin Delta Estuary Hg TMDL study (CEPA, 2005), measurements were made of stormwater runoff during both wet and dry weather periods. In-stream concentrations of T-Hg appeared to higher during wet weather flow while MeHg appeared to be slightly higher during dry weather flows. Concentration of T-Hg on particulates also appeared to be somewhat higher during dry weather. Median and 75th percentile T-Hg concentrations during wet weather flows were about 25 and 50 ng/L respectively. Median and

[11] Note that although many past studies have measured Hg in urban runoff, many of these were not performed using clean techniques, potentially biasing values upward. However, the most significant issue with many of these studies was the high detection limit used (primarily EPA Method 245, whose detection limit of 200 ng/L was too high to accurately measure levels in water). As a result many of the older urban runoff studies show non-detectible levels.

[12] Values shown from Mason and Sullivan are the average of data reported for two branches of the Anacostia River. Values shown from the CEPA study are median values from 12 different sites where measurements were taken.

75th percentile MeHg concentrations were about 0.2 and 0.4 ng/L. Using the median values, %MeHg appears to be about 0.8%. These values appear to be higher in T-Hg than those reported for non-urban runoff (see Table 4-5 below), about the same with regard to MeHg, and therefore appear to have a lower %MeHg than non-urban stormwater. Median and 75th percentile T-Hg concentrations in suspended solids were about 0.4 μg/g and 0.8 μg/g, with values ranging up to 14.1 μg/g.

Mason and Sullivan (1998) measured T-Hg, MeHg and other Hg species in two branches of the Anacostia River, an urban river within the greater Washington, DC area over a range of flow regimes. During base flow conditions, total MeHg averaged 0.12 ng/L in the NE branch and 0.07 ng/L in the NW branch and was 1.4% and 1.6% of T-Hg, respectively. During intermediate flows, total MeHg averaged 0.24 ng/L in the NE branch, 0.27 in the NW branch, and was 0.78% and 0.93% of total T-Hg. During high flows, total MeHg was 0.39 ng/L in the NE branch, 0.77 ng/L in the NW branch, and was 1% and 2.5% of total T-Hg. Reactive ($SnCl_2$ reducible) Hg was also determined. Reactive Hg was determined to be 0.97 ng/L, 1.32 ng/L and 1.89 ng/L under low, intermediate, and high flows, respectively in the NE branch and 0.84 ng/L, 1.03 ng/L and 1.39 ng/L under the same flow regimes in the NW branch. The percentage of reactive/total Hg was 11.1% under low flow conditions but dropped to 4.3% and 4.8% under intermediate and high flow conditions, respectively. Similarly, in the NW branch, the % reactive/total Hg under flow flows was higher (30.3%), but lower under intermediate and high lows (3.5 and 4.5%, respectively). It is of interest to note that the reactive Hg pool is on the order of 10 times higher than the MeHg pool under low flow conditions and on the order of 3 times higher during high flows. That the reactive Hg pool is greater than the MeHg pool was noted earlier with regard to Hg mine drainage waters. Effler and Bloom (1990) reported elevated reactive Hg levels in Onandaga Lake, which receives primarily urban runoff, of 0.5 to 10 ng/L. T-Hg levels ranged from 2 to 25 ng/L which suggests reactive Hg on the order of 25 to 40% of total.

Both studies of sediments in urban drainages were from the San Francisco Bay area. Kinnetic Laboratories, Inc. (2002) reported on a joint stormwater agency study of urban sources of Hg in a number of San Francisco Bay Area counties including Santa Clara, Contra Costa, San Mateo, Marin and Solano. Both MeHg and total Hg in sediments were reportedly analyzed; however, MeHg results do not appear to have been reported. T-Hg was normalized to the fine fraction (<63 microns) weight and is reported as a concentration relative to the dry weight of fines. The study found differences in Hg concentration in urban (n=129) and non-urban (n=22) sites. In the urban sites, T-Hg in sediments ranged from 0.07 to 39.6 μg/g. The mean T-Hg concentration was 2.6 μg/g. The 25th, 50th and 75th percentile values were 0.43, 0.93 and 2.35 μg/g. In non-urban sites, T-Hg in sediments ranged from 0.05 to 14 μg/g. The mean T-Hg concentration was 0.92 μg/g. The 25th, 50th and 75th percentile values were 0.09, 0.32 and 0.43 μg/g. Although actual stormwater samples were not taken as part of this project, TSS concentrations for various land use types were summarized from other publications. "Average" TSS for residential, commercial and industrial were reported to be 90, 98, and 157 mg/L while agricultural and open land uses had values of 2068 and 85 mg/L, respectively. The implication of these numbers is that, while T-Hg concentrations in sediments of urban drainages are on the order of three times higher than their non-urban counterparts, T-Hg concentrations in stormwater from agricultural sources could be on the order of six times greater due to the greater TSS concentrations. When comparing different systems, the concentrations on particles have been reported on a per gram basis. As Hg is strong bound to particles, in any system the total Hg concentration will be a strong function of the sediment loading or TSS.

Based on this study, it appears that runoff from urban drainages have different levels of T-Hg than that of predominantly agricultural drainages. In urban drainages in several San Francisco Bay area counties, T-Hg on particles ranged up to about 40 μg/g, averaging about 3 μg/g. These values are similar to sediments of drainages from gold and other metals mining areas. Concentrations in sediments in watercourses draining agricultural sites were lower, ranging up to 14 μg/g with average concentrations of less than 1 μg/g.

Gunther et al. (2001) report T-Hg and MeHg data from sediment sampling of 21 creeks in western Alameda County, CA. T-Hg concentrations ranged from 35 to 800 ng/g (ppb) and MeHg concentrations ranged from 0.03 to 3.9 ng/g. These concentrations are also adjusted to the weight of sediment in the <62 micron size range. Geometric mean concentrations of T-Hg, MeHg and %MeHg were 142 ng/g, 0.64 ng/g and 0.45% respectively. MeHg concentrations appeared to be correlated with TOC% ($r = 0.59$) and percent fines <63 microns ($r = 0.62$). Percent MeHg was not correlated with TOC ($r = 0.18$) but was somewhat correlated with fines <63 microns ($r = 0.41$). T-Hg concentrations in western Alameda County are substantially lower than those reported by Kinnetic Laboratories above for other Bay Area counties; however, MeHg percentages were on the same order as those typically found in sediments of receiving lakes, streams and estuaries.

Thomas et al. (2002) report that TSS in the Guadalupe River, CA ranged in concentration from 0.5 to 4 μgT-Hg/g (dry weight). This river drains the historic New Almaden Hg mining district. This range is also similar to particulate Hg reported in wastewater treatment plant effluents (see Table 3-1) and also similar to values reported in the Sacramento-San Joaquin Hg TMDL. The article by Mason and Sullivan reported below was the only reference the researchers found to MeHg and R-Hg in urban runoff.

Internationally, Ordonez et al. (2003) measured an average of 2.56 μg/g T-Hg in 112 samples of street dust in Aviles in northern Spain, an industrial city of about 80,000. In Mieres, Spain, a town of 27,000 where the environment has been impacted by historical mining and metallurgical activities, T-Hg in street dusts averaged 4.24 μg/g (Loredo et al., 2003).

Although data is limited for T-Hg and MeHg aqueous concentrations in urban runoff studies, they can be estimated using data found in some of those studies. For instance, Kinnetic Laboratories, Inc. report sediment concentrations in urban drainages adjusted to fines (<63 microns). Conaway et al. (2003) found that T-Hg concentrations in suspended sediments in San Francisco Bay were very similar to sediment concentrations when adjusted to fines <63 microns. Using the fines adjusted T-Hg means in sediments found in the Kinnetic Laboratories study for the urban (2.6 μg/g) and a mean TSS concentration for residential/commercial/industrial areas (115 mg/L) an estimate of 299 ng/L for T-Hg in runoff would be estimated. Using a "low" mean TSS concentration (36 mg/L), the T-Hg concentration estimate would be 93 ng/L. Both of these estimates seem high based on the concentration ranges given in Table 4-6. Using the average and "high" TSS concentrations with the T-Hg concentration reported by Gunther et al. above would give estimated T-Hg in runoff of 16 and 52 ng/L. These estimates are more in keeping with the ranges reported in Table 4-6. The difference is that the means reported in the Kinnetic Laboratory study are 6.5 to 20 times higher than the mean reported in the study by Gunther.

Bonzongo and Lyons (2004) reported a sample collected from the Mobile River as it flows through the City of Mobile, AL to contain 26 ng/L T-Hg and 0.283 ng/L MeHg (1.1%MeHg). This T-Hg concentration is in contrast to all other samples collected from the river which ranged between 0.23 to 5.83 ng/L. While the results of one data point are far from conclusive, it shows the potential for higher levels of Hg in urban runoff.

4.2.5 Non-Urban Runoff

Non-urban runoff represents an ever-present and significant Hg source in nearly all watersheds. A number of studies were found in the literature reporting levels of various species of Hg in non-urban runoff. Those reviewed as a part of this study are summarized in Table 4-7. With the exception of the East Fork Poplar Creek, TN, a waterway contaminated by Hg released from the production of nuclear weapons manufacture, representative T-Hg levels appear to be less than 20 ng/L[13], although values of up to 80 ng/L are reported for rivers with high TSS. Dissolved T-Hg can represent a substantial and quite variable fraction of total Hg in non-urban runoff, in these studies ranging from <10% to about 60%. Higher percentages of dissolved T-Hg will typically be associated with systems having lower TSS or under baseflow conditions in any stream when TSS is typically lower. MeHg appears to range up to a concentration of 0.3 to 0.4 ng/L, except in some systems where DOC levels would appear to sustain them at higher levels (e.g. in the blackwater streams of GA and wetland/peatland areas). Percent MeHg appears to be higher in non-urban runoff (typically 5-15%) than in other nonpoint sources (urban runoff: 1-2%; mine drainage: < 1%, and precipitation: 1-5%).

In a review paper, Grigal (2002) documents the range of T-Hg in streamflow from terrestrial watersheds is on the order of 1.2 to 20 ng/L (n=79). On the basis of comparison of regression relationships between T-Hg and MeHg fluxes from a number of watersheds, he suggests that the upper limit on %MeHg in streamflow from terrestrial watersheds may be on the order of 10 to 15%, a figure in line with the observations in Table 4-7.

For comparative purposes, some data from North American and Swedish lakes are also shown in Table 4-7. Mason and Sveinsdottir (2003) measured total and dissolved T-Hg and total and dissolved MeHg in 18 Lakes in Maryland. Most of the lakes were described as having watersheds predominantly in forest or agricultural land uses and many of them are water supply reservoirs. Average dissolved T-Hg was reported to range from 0.4 to 19.5 ng/L with an overall mean of 3.6 ng/L. Average dissolved MeHg concentrations were reported to range between 0.016 and 0.292 ng/L with an overall mean of 0.131 ng/L. Based on the averages reported, the %MeHg was 3.6%. Based on 42 values, the average dissolved/total T-Hg for these lakes was 87%. However, this figure may be somewhat high since several of the filtered concentrations were reported as exceeding the unfiltered concentrations.

Shanley et al., 2005 studied data from 219 lakes and 474 streams in New York, New England, and Eastern Canada for the purpose of deriving predictive models for T-Hg and MeHg in streams and lakes in this area. The database contained 693 T-Hg observations and 385 MeHg observations. For streams, the median MeHg concentration was about 0.1 ng/L and the median ratio of MeHg/T-Hg was about 5%. The 75th percentile values were about 0.2ng/L and about 10%. For lakes, the median MeHg concentration was about 0.25 ng/L and the median MeHg/T-Hg ratio was about 15%. The 75th percentile value for lakes was about 0.4 ng/L and the MeHg/T-Hg ratio about 20%. Rolfhus et al. (2003) reported on Hg speciation in Lake Superior. The

---

[13] Note that in Table 4-6, single numbers represent average of media values while ranges indicate minima and maxima, except where noted.

Table 4-7. Representative Concentrations of T-Hg, MeHg and %MeHg for Runoff and Streamflow of Primarily Non-Urban Areas Reported in the Literature.

	Unfiltered T-Hg (ng/L)	Dissolved T-Hg (ng/L)	% Diss. T-Hg (ng/L)	MeHg (ng/L)	%MeHg	Reference
East Fork Poplar Creek, TN/U.S.	589	197	33%	0.332 (Unfiltered) 0.157 (Filtered)	0.06% (Unfiltered) 0.08% (Filtered)	Reported in EPRI (2003)
Six Blackwater Rivers, Georgia/U.S.	1 to 4 (Fall) 5 to 15 (Spring)			0.3 to 1 (Fall) 0.3 to 3 (Spring)	5 to 35% (Fall) 5 to 45% (Spring)	EPA (2001)
Minnesota River, MN/U.S.	0 to 20		<10%			Balogh et al. (1998)
Headwaters Mississippi River, MN/U.S.	0 to 5		28%			Balogh et al. (1998)
St. Croix River	0 to 2		62%			Balogh et al. (1998)
Minnesota, Blue Earth, Le Sueur River, south central MN/U.S.	up to 80			< 1	1.7 to 12.2%	Balogh et al. (2003)
Six rivers, WI/U.S.	0.26 to 45.9			< 0.01 to 1.77	0.69 to 9.7	Babiarz et al. (1998)
New England Streams/U.S.	0.1				5%	Shanley et al. (2005)
Trott Brook, Clear Creek, East-central Minnesota/U.S.	0.8 to 2.0 (Baseflow) 2.0 to 3.5 (Stormflow)	0.2 to 1.0 (Baseflow) 1.0 to 2.8 (Stormflow)	> 65%	0.1 to 0.2 (Baseflow) 0.2 to 1.5 (Stormflow)	0.05 to 0.17% (Baseflow) 0.34 to 0.44 % (Stormflow)	Balogh et al. (2004)
Northern Sweden	4.85			0.24	4.9%	Bernhardsson (2005)
Northern Sweden	3.64			0.42	11.5%	Bernhardsson (2005)
Seven forested catchments, Sweden	2 to 12			0.04 to 0.64	3.48%	Lee and Iverfeldt (1991)
Eight forested catchments, Finland	0.84 to 24 3.7 to 6 (Range of medians)			0.03 to 3.8 0.16 to 0.88 (Range of medians)	0.34 to 56% 2.8 to 28% (range of medians)	Porvari and Verta (2003)

Table 4-7. Representative Concentrations of T-Hg, MeHg and %MeHg for Runoff and Streamflow of Primarily Non-Urban Areas Reported in the Literature.

	Unfiltered T-Hg (ng/L)	Dissolved T-Hg (ng/L)	% Diss. T-Hg (ng/L)	MeHg (ng/L)	%MeHg	Reference
Minnesota/U.S.	20.6 (Upland) 12.9 (Wetland)					Reported in Porvari and Verta (2003)
Wisconsin/U.S.	0.9 to 5.9 (Wetland)			0.08 to 0.21 (Wetland)		Reported in Porvari and Verta (2003)
Adirondack Mtns., NY/U.S.	2.3 (Wetland)			0.10 to 0.40 (Wetland)		Reported in Porvari and Verta (2003)
ELA, Ontario, Canada	8.8 to 13 (Upland) 3.5 to 4.8 (Wetland)			0.031 to 0.055 (Upland) 0.46 to 0.73 (Wetland)	0.23% (Upland) 13% (Wetland)	Reported in Porvari and Verta (2003)
San Francisco Bay, CA/U.S.	0.15 to 88	0.08 to 34	10% to 20%	<0.01 to 0.5	0.64%	Conaway et al. (2003)
Mobile River, AL/USA	2.14	0.40	18.7%	0.063	2.9%	Bonzongo and Lyons (2004)
Mobile Bay, AL/U.S.	1.33	0.43	32.3%	0.050	3.8%	Bonzongo and Lyons (2004)
North American and Swedish Lakes						
New England Lakes/U.S.	0.25				15%	Shanley et al. (2005)
Maryland Lakes, MD/U.S.		0.4 to 19.5	87%	0.016 to 0.292	3.6%	Mason and Sveinsdottir (2003)
Various lakes and impoundments, northwestern MN/U.S.	0.37 to 13.7	0.69 to 7.75		0.014 to 8.36 (Unfiltered) 0.024 to 3.63 (Filtered)	0.9 to 94%	Brigham et al. (2002)
Lake Superior, North America	0.47			0.0058	1.2%	Rolfhus et al. (2003)
Eight drainage lakes, Sweden	1.35 to 15			0.04 to 0.8	1 to 12%	Lee and Iverfeldt (1991)

average concentration of unfiltered T-Hg was quite low (0.47 ng/L). Dissolved MeHg was 0.0058 ng/L or 1.2% of total T-Hg. Reactive ($SnCl_2$ reducible Hg) was nearly an order of magnitude greater than MeHg, with an average concentration of 0.051 ng/L or 10.8% of unfiltered T-Hg. Brigham et al. 2002 reported MeHg and T-Hg in filtered and unfiltered samples from various types of lakes in northwestern MN including a relatively new impoundment, older permanent pool impoundments, temporary pool impoundments and reference lakes. Unfiltered T-Hg ranged from 0.37 to 13.7 ng/L over all lakes with filtered T-Hg ranging from 0.69 to 7.75 ng/L. Unfiltered MeHg ranged from 0.014 to 8.36 ng/L and %MeHg (unfiltered) ranged from 0.9 to 94%. Lee and Iverfeldt (1991) sampled lake water from eight drainage lakes. Resulting T-Hg concentrations ranged from 1.35 to 15 ng/L and MeHg concentrations ranges from 0.04 to 0.8 ng/L with %MeHg ranging from 1 to 12%.

T-Hg in lakes appears for the most part to be lower than runoff or stream values and dissolved percentages may be higher. This is likely due to more quiescent environment in lakes conducive to settling of particulates. In such environments, T-Hg values would be expected to be lower and % dissolved T-Hg to be higher.

Data for the East Fork Poplar Creek, TN in the vicinity of the Oak Ridge National Laboratory (as reported in EPRI, 2003) show that T-Hg concentrations had a mean of 589 ng/L in unfiltered samples and 197 ng/L in filtered samples. Thus the ratio of dissolved to total T-Hg for this system is on the order of 33%. The mean value reported for total MeHg was 0.332 ng/L with the dissolved MeHg reported as 0.157; yielding a ratio of dissolved to total MeHg of about 47%. Based on these data, relative MeHg total is estimated to be 0.06% for unfiltered samples and 0.08% for filtered samples.

In 2000 and 2001, U.S. EPA (U.S. EPA, 2001) measured Hg and ancillary water quality parameters in six blackwater systems in South Georgia as part of the TMDL effort. In a limited number of samples from each system, unfiltered T-Hg ranged from about 1 ng/L to 4 ng/L in the fall of 2000, during a time of low flow under drought conditions. In the spring of 2001, under high flow conditions, unfiltered T-Hg ranged from about 4 ng/L to 17 ng/L in these systems. Land use in all of these watersheds was primarily in forest and agricultural production with limited urban development. Unfiltered MeHg in the water column ranged from about 0.3 to 1 ng/L in 2000 and from 0.1 to 2.8 ng/L in 2001. Relative MeHg concentrations ranged from about 5% to 35% in the fall of 2000 and from 2% to 25% in the spring of 2001.

Studies in MD found an inverse relationship between %MeHg and flow for the Potomac and Susquehanna Rivers, which are both large systems (Mason et al., 1999; Lawson et al., 2001) (Figure 4-4). For other smaller rivers in MD, the relationship was not as clear, but there was strong evidence for an increase in %MeHg with increasing flow. In most cases, however, while the total Hg concentration was strongly correlated with flow rate, this was not so for MeHg. These relationships suggest that during high flow periods, resuspension or bank erosion contributes particles to the system that have a lower %MeHg than those present under baseflow conditions. This likely reflects the notion that under baseflow conditions, a higher fraction of the particles are biotic (plankton) and which are known to accumulate Hg, and especially MeHg, to higher levels than inorganic particles. This trend is also apparent, for example, during the resuspension studies of Kim et al. (2006) where the %MeHg on particles was much higher in the mesocosms without resuspension compared to those with resuspension; these particles also had a

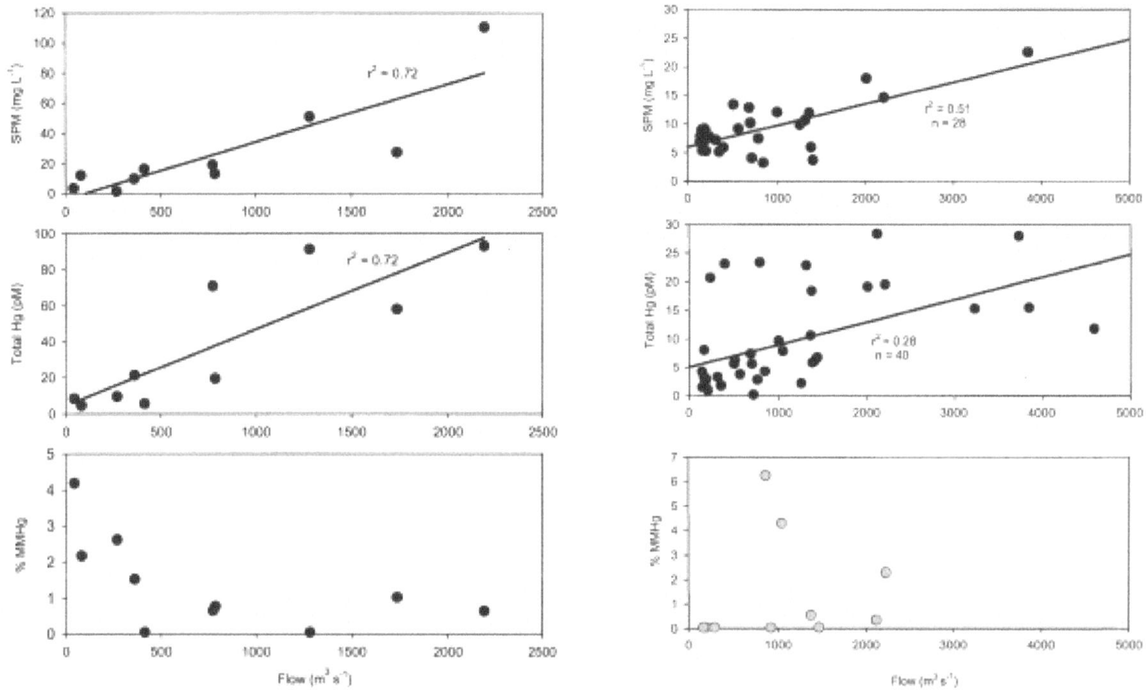

Figure 4-4. Relationship between Total Mercury, Methylmercury and Flow Rate for the Potomac (left) and Susquehanna Rivers (right). Reprinted with permission from Lawson et al., *Water Research*, Elsevier (2001).

higher percentage of particulate organic carbon (%POC). Resuspension in these mesocosm studies using contaminated Baltimore Harbor sediment, substantially increased TSS and total Hg concentration per liter but did not substantially increase the dissolved Hg concentration. For MeHg, resuspension decreased the MeHg concentration per gram on particles in the water column. Thus, for dynamic systems, it is possible that the %MeHg within the system can change over time due to the differences in suspended solid %MeHg under baseflow conditions (a few percent) to those of high flow or high TSS concentrations (<1%MeHg).

Balogh et al. (1998) report Hg, flow and TSS data for three large non-urban drainages (the Minnesota River, headwaters of the Mississippi River, and St. Croix Rivers) in Minnesota over a two year period. T-Hg concentrations in the Minnesota River ranged primarily between 0 and 20 ng/L with TSS concentrations in the 0 to 500 mg/L range. In the headwaters of the Mississippi River, T-Hg concentrations were primarily in the 0 to 5 ng/L range, with TSS in the 0 to 40 mg/L range. In the St. Croix River, T-Hg concentrations were mostly in the 0 to 2 ng/L range, with

TSS ranging from 2 to 8 mg/L. Dissolved T-Hg concentrations as a percentage of unfiltered T-Hg were reported to be 62% in the St. Croix, 28% in the headwaters of the Mississippi, and less than 10% in the Minnesota River. Although the authors make this point, % dissolved T-Hg appears to be higher in these three rivers when the T-Hg concentrations are lower. T-Hg in the St. Croix River was found not to be correlated with TSS ($R^2 = 0.04$), whereas in the Minnesota River there was a strong relationship ($R^2 = 0.97$). This led the authors to conclude that the dissolved phase dominates Hg transport in the St. Croix and that DOC may play a more dominant role in Hg transport than TSS. Of the three rivers, the St. Croix has the highest percentage of forested land and thus sediment loads may be naturally lower.

Balogh et al. (2003) reported MeHg and T-Hg concentrations in runoff of the Minnesota, Blue Earth, and Le Sueur River, draining predominantly agricultural watersheds in south-central Minnesota. T-Hg concentrations raged up to 80 ng/L and were highly correlated with TSS. MeHg concentrations were generally less than 1 ng/L with the bulk of the values being less than < 0.6 ng/L. MeHg concentrations were also correlated with TSS, although less strongly than T-Hg. Percent MeHg for these systems were calculated from the slopes of the T-Hg and MeHg regressions with TSS. Values were calculated to be between 2.4% and 4.2% for the two locations on the Minnesota River, 12.2% for the Blue Earth River, and 1.7% for the Le Sueur River.

In another study in North-central Minnesota, Balogh et al. (2004) reported both dissolved and total T-Hg, MeHg and %MeHg for two small (193-194 km^2) watersheds draining primarily agricultural (40 to 50%) and wetland (9 to 10%) land uses. Sampling occurred approximately weekly between late April and late October. Unfiltered T-Hg ranged between 0.8 and 2 ng/L during baseflow periods but increased to the range of 2.0 to 3.5 during a storm that occurred in late June. Filtered T-Hg was reported to be > 65% in the filtered (dissolved) phase and > 82% of the MeHg was reported to be in the dissolved phase during the storm event. Unfiltered MeHg concentrations ranged between 0.1 and 0.2 ng/L during baseflow and 0.2 to 1.5 ng/L during stormflow. Percent MeHg was lower during baseflow periods (0.05 to 0.17%) and higher during stormflow periods (0.34 to 0.44%).

Babiarz et al. (1998) measured T-Hg and MeHg in rivers draining watersheds having a variety of land uses. T-Hg ranged from 0.26 to 45.9 ng/L while MeHg ranged from < 0.01 to 1.77 ng/L. The highest T-Hg concentration was associated with a predominantly agricultural watershed while the highest MeHg concentration was associated with a watershed dominated by wetlands. Mean event-based concentrations ranged from 1.21 to 8.05 ng/L. Means of non-event based samples exhibited similar T-Hg ranges (1.25 to 7.95 ng/L). Interestingly, the predominately wetland watershed exhibited the highest average T-Hg concentrations (higher than the predominately agricultural watershed). A predominately forested watershed exhibited the lowest mean T-Hg concentrations. The predominately wetland watershed also exhibited the highest MeHg concentrations and the forested watershed the lowest MeHg concentrations (excepting the urban watershed). Percent MeHg concentrations calculated from the means of the event and non-event concentrations ranged between 0.69 and 9.7% with only minor differences between event and non-event percentages. In this study, $\log K_D$ was found to range between 2.8 and 5.5 for T-Hg and 2.9 and 5.8 for MeHg. These partition coefficients were compared with those from other studies (i.e. 5 to 6.9 for Lake Michigan tributaries, 4.5 to 5.7 for Wisconsin lakes, and 5.5 to 5.7, open waters of Lake Michigan) and were found to be comparable.

Sloan et al., 2001) measured total T-Hg and dissolved T-Hg from a watershed on which Hg containing biosolids had been applied. Unfiltered T-Hg concentrations from the control plot ranged from 9.2 to 27.9 ng/L and averaged 17.7ng/L on 4 consecutive days with snowmelt runoff. Unfiltered T-Hg in runoff declined each day on which measurements were made indicating that perhaps the supply of Hg in sediments was limited. Filtered T-Hg samples ranged from less than detection (detection level not given) to 4.7 ng/L. The average of the samples in which dissolved Hg was detected was 3.8 ng/L. No declining trend was observed in filtered T-Hg concentrations as it was for unfiltered T-Hg. Percent dissolved T-Hg ranged from 13.7% to 51%. It is of interest to note that the surficial T-Hg concentrations in the soil in the control ranged from about 30 to 50 μg/g. If you take the average soil T-Hg concentration and divide it by the dissolved T-Hg concentration in runoff, a partition coefficient results with a value ($logK_D$) of 7.0.

In the Mobile River system, Alabama, Bonzongo and Lyons (2004) reported T-Hg in 31 samples of inland waters to range between 0.23 and 5.83 ng/L with a median of 2.14 ng/L. Dissolved concentrations ranged from 0.04 to 2.61 ng/L with a median concentration of 0.4 ng/L. Total and dissolved MeHg were also measured. Total MeHg ranged from below detection to 0.938 ng/L with a median of 0.063 ng/L and dissolved MeHg ranged from below detection to 0.38 ng/L, with a median of 0.036 ng/L. Based on median concentrations, the dissolved to total ratio was 18.7% for T-Hg and 57% for MeHg. Measurements (n=5) were also made in Mobile Bay. T-Hg concentrations were lower (median of 1.33 ng/L unfiltered) whereas filtered T-Hg was slightly higher than for the inland waters (0.43 ng/L), yielding a % dissolved T-Hg of 32%. MeHg concentrations were also lower (median of 0.056 ng/L unfiltered) and filtered T-Hg was slightly higher than for the inland waters (0.037 ng/L), yielding a % dissolved T-Hg of 66%. Based on median values, %MeHg on unfiltered samples was 3.8%.

In northern Europe, Bernhardsson (2005) reported MeHg and T-Hg concentrations in runoff from two forested catchments in northern Sweden. MeHg in the Balsjo catchment ranged from 0.09 to 0.49 ng/L with an average of 0.24. T-Hg concentrations in runoff 1.78 to 13.1 ng/L with an average of 4.85 ng/L. In the Vindeln catchment, MeHg in runoff ranged from 20.1 to 0.6 ng/L with an average of 0.42 ng/L and T-Hg ranged from 2.5 to 5 with an average of 3.64 ng/L. Based on average values, the %MeHg in runoff from these watersheds were 4.9% and 11.53% respectively. Soils in these studies are referred to by the author as "mires" and would tend to be high in organic matter; which may partially explain the elevated MeHg levels.

Porvari and Verta (2003) measured MeHg and T-Hg in runoff from eight small boreal forest catchments during 1990 to 1995 in Finland. Soils included both mineral and peatlands. MeHg in runoff ranged from 0.03 to 3.8 ng/L and T-Hg ranged from 0.84 to 24 ng/L. The ranges of the medians for each catchment were 3.7 to 6 ng/L for T-Hg and 0.16 to 0.88 ng/L for MeHg. Percent MeHg ranged from 0.34 to 56%. Range of the median values for each catchment was 2.8 to 28%. T-Hg in runoff from catchments with primarily mineral soils (5.8 ng/L) were slightly higher than catchments with primarily peat soils (4.8 ng/L). However, concentrations of MeHg were dramatically higher from catchments with peat soils (0.77 ng/L) than from those with mineral soils (0.18 ng/L) with correspondingly higher %MeHg ratios (19% for peat versus 3.9 for mineral soils). Concentrations of T-Hg and MeHg from these catchments were compared to concentrations measured in other studies worldwide. In general, T-Hg concentrations were intermediate in this study compared to others. Reported mean concentrations for waters draining upland and peatland soils in Minnesota were 20.6 ng/L and 12.9 ng/L, respectively. Reported mean T-Hg concentrations from Wisconsin watersheds ranged from 2.73 ng/L to 3.92 ng/L and MeHg was reported to range from 0.08 to 0.21 ng/L from a wetland watershed (T-Hg 0.9 to 5.9 ng/L). T-Hg concentrations from a wetland watershed in the Adirondack Mountains, NY was 2.3 ng/L with MeHg ranging from 0.10 to 0.40 ng/L. Mean T-Hg in waters from upland and wetland watersheds in the Experimental Lakes Area, Ontario, Canada, ranged from 3.5 to 13 ng/L with MeHg ranging from 0.031 to 0.055 ng/L for an upland watershed to 0.46 to 0.73 ng/L for a wetland watershed. Percent MeHg in the ELA watersheds ranged from 0.23% for the upland to 13% for the wetland watershed.

For comparative purposes, the researchers included speciation data from San Francisco Bay in Table 4-7. Even though Bay waters are influenced by urban sources around its margin, there are major non-urban components to the inflows and, in addition, the inflows are influenced by historic gold and Hg mining activities. Another difference between the other entries in the table and the San Francisco Bay is that its waters are estuarine, whereas the others are fresh. Conaway et al. (2003) report data that were collected in various seasons over several years. Unfiltered T-Hg concentrations of 0.15 to 88 ng/L cover the range of the riverine systems in the table, with the exception of the contaminated East Fork Poplar Creek. The same holds for the unfiltered T-Hg concentrations. The percentage of dissolved T-Hg appears to be at the low end of those reported for riverine systems. MeHg concentrations appear to be somewhat lower than the reported riverine systems but the %MeHg appears to be within the same range. Given that % dissolved T-Hg and %MeHg are lower than those reported for riverine systems, the bioavailability of Hg in the San Francisco Bay estuary would appear to be lower than that in fresh waters.

4.2.6 Wastewater Treatment Plants

Wastewater treatment plants are the only point source profiled as part of this study. As noted in earlier chapters, there in a dearth of information related to these sources. Most of the relevant studies were reviewed in Chapter 3.0; however, a few additional studies are especially noted here. Some data collected by EPA Region 4 as part of its Hg TMDL investigations in Georgia showed MeHg levels ranging from about 0.01 to 0.68 ng/L. Percent MeHg ranged from about 1.5 to 7%. Data from WWTPs in the Sacramento area as reported by Abu-Saba et al. (2005) document MeHg levels in WWTP discharge of 0.02 to 2.6 ng/L. Data from the Western Lake Superior Sanitation District (WLSSD unpublished) show an average MeHg in its effluent of 0.35 ng/L with %MeHg of 11%. All these data appear to be of similar magnitude to those reported in Table 3-2. Further details of these studies are provided below.

A few measurements of MeHg in WWTP effluent were made as part of EPA Region 4's Hg TMDL efforts in Georgia. Data were found for four WWTPs in US EPA (2001a). Measurements in the Athens-North Oconee WWTP (11 MGD) revealed MeHg of 0.68 ng/L and T-Hg of 31 (2.2%MeHg). In the Polebridge Creek WWTP (20 MGD), MeHg of 0.06 ng/L versus T-Hg of 0.86 (both values estimated) yielded 7%MeHg. The Snapfinger Creek WWTP (36 MGD) had effluent levels (estimated) of 1.7 ng/L T-Hg and 0.03 ng/L MeHg yielding 1.8%MeHg. Finally, the Camp Creek WWTP (1.5 MGD) had effluent T-Hg of 0.73 ng/L and MeHg of 0.01 MeHg (values estimated), a 1.4%MeHg ratio. A duplicate sample from the Camp Creek plant showed T-Hg at 1.2 ng/L and MeHg at 0.02 ng/L (values estimated) for a methyl/total ratio of 1.7%.

A regional study of WWTPs in the Sacramento River watershed was performed in 2004/05 (Abu-Saba et al., 2005). Although all of the raw data are not contained in this report, the data in the report do indicate MeHg concentrations ranging from 0.02 to 2.6 ng/L. The highest values are from one plant (City of Stockton) with an average (n=2) effluent concentration of 2.2 ng/L. The plant with the highest number of observations (n=5 over a 7 month period) had an average MeHg effluent concentration of 0.24 ng/L with a standard deviation of 0.15 ng/L (CV = 64%).

The longest term record available to us for MeHg and T-Hg in WWTP effluent was provided by the Western Lake Superior Sanitation District (WLSSD) (Tuominen, 2006 personal communication). This data show that the average T-Hg concentration in its WWTP effluent from

2002 to 2006 (n=121) was 2.63 ng/L with a standard deviation of 0.16 ng/L (CV = 6.1%). The minimum and maximum values in the data set were 1.2 ng/L and 13.6 ng/L respectively. MeHg was measured only three times and averaged 0.353. Based on these three data points, the ratio of methyl to total Hg was 11% of the average total T-Hg in effluent of 3.2 ng/L. In addition, filtered T-Hg was 54% of unfiltered T-Hg for these three data points. In 2006-2007, dissolved and total T-Hg were measured in the plant effluent. Based on these measurements (n=20), average unfiltered T-Hg was 2.78 ng/L and dissolved T-Hg was 1.47 ng/L. The average of the dissolved/total ratios was 55%. The average TSS concentration for the same data averaged 4.61 mg/L. Therefore, the average concentration in the suspended particles was 0.28 ng/mg or µg/g. Thus, the K_D for the partitioning of T-Hg to solids was 1.93E5 L/kg ($\log K_D = 5.3$). By comparison, the average $\log K_D$ for T-Hg measured in the fluvial portion of the Scheldt River by Leermakers et al. (2001) was about 5.5 (by contrast, the average $\log K_D$ for MeHg was about 5.)

Bodaly et al. (2004) reported MeHg levels from unfiltered samples in the effluent of three wastewater treatment plants in the City of Winnipeg, Manitoba, Canada. MeHg levels in treated sewage effluent ranged from 0.1 to 0.4 ng/L. Comparison to MeHg levels in raw sewage (0.5 to 4.3 ng/L) suggests removals on the order of 80% to 91%. MeHg levels in treated effluent were reported to be similar to levels in the receiving waters and therefore the effect of the discharged effluent was to only slight change MeHg levels in the receiving waters (usually a change of 2% or less). In one case, however, discharge of up to 2 ng/L MeHg from the West End plant into the Assiniboine River resulted in increases in the receiving water of up to 11 %.

As reported in Pavlogeorgatos (2006), Goldstone studied MeHg concentrations in a full-scale WWTP in Norwich and reported <0.5% of the input to the plant was MeHg. However, detection limits (10 ng/L) were probably not adequate to determine MeHg in effluents. Pavlogeorgatos (2006) also found no MeHg in a pilot-scale activated sludge WWTP supplied with synthetic wastewater and enriched with Hg. Again, however, the detection limit achieved in wastewater samples (70 ng/L) was probably not adequate to detect MeHg at levels that might have been present.

Gilmour and Bloom (2005) studied T-Hg and MeHg in a 40 MGD WWTP in which elemental Hg was used as a seal in three trickling filter columns. Effluent MeHg concentrations were reported to be on the order of 4 to 6 ng/L. It is not known how commonplace the use of such seals is in modern WWTPs; however, the levels of MeHg reported in the effluent are an order of magnitude greater than the levels reported by Bodaly et al. (2004) of 0.1 to 0.4 ng/L and greater than the range shown in Chapter 3.0, Table 3-2, of this report (<0.1 to 1.69 ng/L; average 0.44 ng/L, standard deviation 0.53 ng/L) for 20 WWTPs in the U.S.

4.3 Enhancer/Mitigator Source Profiles

In this section the sources identified above are profiled with regard to those factors in the effluent that may enhance or mitigate the bioavailability of Hg. To recapitulate the conclusions of Chapter 3.0, the factors of greatest importance, other than those directly related to Hg or MeHg, are the TSS concentration, the strength by which Hg is bound to solids (as suggested by the partition coefficient), and the organic content of water and/or solids (as measured by volatile suspended solids (VSS), dissolved organic carbon (DOC) or NOM, particulate organic carbon (POC), and or total organic carbon (TOC)). Secondarily, sulfate is a helpful parameter as it has been shown in certain instances to be capable of stimulating MeHg production. Finally, pH, conductivity and chloride levels may be helpful in certain special instances[14].

Table 4-8 below attempts to characterize the concentrations of Hg methylation enhancers and mitigators found in the literature. Several facets of these entries are worth noting. First, the comparisons are presented as "order-of-magnitude" estimates. This was done after attempts to characterize the observations in other ways (means, ranges) provided a less meaningful assessment. In some cases, ranges are presented. The setting of these ranges involved the judgment of the authors. In some cases, the ranges encompass the range of information reported in the literature. In some cases, the upper end of the range may be less than the highest observation in the literature, if in the opinion of the authors, the next greater order-of-magnitude was too high[15]. In some cases, single numbers are presented. This may be for two reasons; either the range observed is too small to justify an order-of-magnitude range (e.g. TSS in rainfall) or there were too few observations to justify a range (e.g. chlorides in WWTP effluent).

Other factors may also be important, as identified in Chapter 3.0. However, in many cases, the literature is inconsistent or data have not been collected in sufficient detail or quantity for trends to be assessed. A case in point is nutrients. For instance, Balogh et al. (2003) suggested that elevated levels indicators of biological activity (e.g. chlorophyll, volatile solids and Kjeldahl nitrogen) might signal increases in MeHg concentrations in their study of rivers draining agricultural watersheds in MN. On the other hand, Bonzonga and Lyons (2004) found no connection between nitrates or phosphorus and MeHg abundance in the Mobile River, AL. Vaithiyanathan et al. (1996) found no evidence that nutrient-enriched agricultural runoff contributed significantly to T-Hg buildup in the northern Everglades or that eutrophication affected MeHg production in Everglades peat soils. While connections have been postulated between eutrophication and higher MeHg levels in aquatic ecosystems as a result of increased productivity and higher oxygen demand, leading to enhanced MeHg production, there seems to be no strong support for this effect in the literature. On the other hand, there are citations in the literature suggesting biodilution effects of increased productivity (Chen and Folt, 2005).

4.3.1 Atmospheric Deposition

TSS/VSS: In precipitation as a rule, you would not expect very high levels of particulates. Novotny and Chesters (1981) report levels as being in the range of 11 to 13 mg/L. Although TSS levels were not reported in Lamborg et al. (1995), comparison of filtered versus unfiltered samples showed that 76% of the T-Hg in rainfall samples were associated with particulates using a 0.2 micron filter and 49% of T-Hg in snowmelt samples were associated with the particulate phase.

DOC/POC/TOC: No levels of TOC or DOC in precipitation were found in the literature; however, Novotny and Chesters (1981) report that the COD of precipitation is in the range of 9 to 16 mg/L. Assuming that TOC can be calculated from the stoichiometric ratio of COD/TOC to

---

[14] The researchers speculate that pH may be helpful in situations where higher pH effluents may be discharging into low pH receiving waters. Conductivity may be useful in situations to estimate the effects of particle disintegration/coagulation on bioavailability. Chloride levels may be useful in cases where very high chloride levels (e.g. resulting from the presence of deicing salts in snowmelt) may exist.

[15] For instance, in the case of sulfate in rainfall, observations higher than 1 mg/L (e.g. 3 mg/L) were found but this did not seem to justify extending the upper end of the range to 10 mg/L.

approximate the molecular ratio of oxygen to carbon (i.e. 32/12 or 2.66), TOC may be calculated from COD by multiplying COD levels by 1/2.66 or 0.376. Using the range presented above, this would give TOC of precipitation levels of 3.4 to 6 mg/L.

Table 4-8. Characteristic Concentrations of Methylation Enhancers/Mitigators for Various Mercury Sources (Order-of-Magnitude Comparisons).

	TSS	DOC/TOC	SO4	Cl	pH	Conductivity
Source	**mg/L**	**mg/L**	**mg/L**	**mg/L**	**SU**	**uS/cm**
Atmospheric Deposition	10	1 to 10	0.1 to 1	0.1 to 1	4 - 5	10 to 100
Mining Runoff	100 to 1,000	1		10 to 100	3 - 9	100 to 1,000
- Mercury mines			1 to 10,000			
- Gold/other metals mines			1 to 1,000			
Sediment Pore Waters	Not Applicable					Not Known
- Fresh		10 to 100	10 to 100	1 to 100		
- Saline			100 to 1,000	100 to 10,000		
Urban Runoff	10 to 1,000	1 to 10	10 to 100	10 to 100 1,000 to 10,000 [a]	7 - 9	100 to 1,000
Non-urban Runoff		1 to 10		1 to 100		
- Agricultural	100 to 1000		1 to 100		5 - 8	1000
- Silvicultural	10 to 100		1 to 100		5 - 8	100
- Wetlands	10 to 100		1 to 10		4 - 5	100
Wastewater Treatment Plants	10 to 100 (Secondary) 1 to 10 (Advanced)	10 to 100 (Secondary) 1 to 10 (Advanced)	100	100	6 - 8	1000

[a] Concentrations in highway runoff affected by deicing salts may fall in this range.

Sulfate: Lynch et al. (1996) reported sulfate concentrations at 58 National Atmospheric Deposition Network (NADP) sites. In 1995, sulfate concentrations ranged from a low of 0.15 mg/L at a site in Oregon to 2.78 mg/L at a site in Indiana. Junge and Werby (1958) reported sulfate concentrations in precipitation in the conterminous U.S. ranging narrowly between 1 and 3 mg/L. Long term trend show declines in atmospheric sulfate as controls on sulfur emissions from fossil fuel burning power plants has been curtailed (Lehmann et al., 2007). During the period from 1985 to 2004, mean sulfate concentrations have decreased by over 50% in areas of the western and northeastern US. Decreases from 25 to 50% have been observed over most of the remainder of the country. The lowest decreases have been in Texas, Louisiana, Oklahoma, and Florida.

Chloride: Junge and Werby (1958) reported that Cl concentrations in the conterminous U.S. ranged from around 0.1 mg/L over much of the Midwest to between 1 and 2 mg/L on the west coast and between 1 to 4 mg/L on the east and gulf coasts. Most of the U.S. is between 0.1 and 0.4 mg/L. Measurements between 1955 and 1956. Lynch et al. (1996) reported chloride concentrations ranging from a low of 0.04 mg/L at a site in Montana to 1.14 mg/L at a site in Florida.

pH: Cogbill and Likens (1974) reported pH value in the northeastern U.S. show consistent pH of less than 4.4, rather than the 5.6 for precipitation in equilibrium with atmospheric CO_2. Galloway et al. (1982) measured pH of precipitation in remote areas of the world. Volume weighted pH was consistently between 4.8 and 5. Hendry and Brezonik (1980) reported volume weighted pH of precipitation in Gainesville, FL to be 4.53. Acidity was due to sulfuric acid (69%) and nitric acid (23%). From these limited observations a conservative estimate of pH in U.S. rainfall would be 4.5 to 5. The lower end of this range might be more appropriate to use in areas with higher sulfate levels. Hong Kong rainwater was reported to be pH 4.2 in the period from 1990 to 1992 (Sequeira and Lung, 1995).

Conductivity: The conductivity of rainwater is low, probably in the 10 to 20 uS/cm range, although in more polluted areas it may be in the 40 to 60 uS/cm range, and values up to 160 uS/cm have been reported in the literature in the beginning of storms (Jonsson and Vonnegut, 1991). The mode of conductivity of Hong Kong rainwater was reported in the range of 15 to 20 uS/cm during the period from 1990 to 1992 (Sequeira and Lung, 1995).

4.3.2 Mining Sources

Information provided below on bioavailability enhancers and mitigators addresses gold and other metals mines as well as Hg mines.

TSS/VSS: Domagalski reported TSS levels in the Sacramento River and associated tributaries draining the historic gold and Hg mining areas of California of from 1 to 3000 mg/L, with the bulk of the measurements being about 300 mg/L. Ganguli et al. (2000) report TSS levels from the New Idria mining site in the Coast Range of California. TSS below the mine ranged from 370 to 660 mg/L. Above the mine TSS concentrations were much lower (1.2 to 2.2 mg/L). They also reported logK_Ds for T-Hg and MeHg. Downstream of the mine, the mean value for T-Hg was 4.6 ± 0.25 (n=3) and the mean for MeHg was 3.6 (n=2). LogK_Ds for waters upstream of the mine were considerably higher, averaging 6.4 (n=2). Rytuba reports that logK_D in mine drainage waters is a function of pH as it is controlled by adsorption onto iron oxyhydroxide. LogK_Ds are rather constant between 4 and 5 and above pH 7.1 drop off dramatically.

DOC/POC/TOC: Ganguli et al. (2000) reported that DOC levels in waters downstream of the New Idria mine were quite low, ranging from 0.96 to 1.4 mg/L. Upstream waters were similar, ranging from 0.46 to 1.3 mg/L.

Sulfate: Ashley et al. (2002) reported sulfate levels in waters proximal to an abandoned placer gold mine in Shasta County, CA. Sulfate content of these waters ranged from < 1.6 to 24 mg/L and were reported to be in the normal range seen in West Coast Rivers. Rytuba (2002) reports that sulfate concentrations exceed several thousand mg/L from silica-carbonate deposits with high iron sulfide content and shows a range of concentrations from about 3 to over 10,000 mg/L. Rytuba (2000) reported sulfate data from both silica-carbonate and hot-spring type mines in the California Coast Range. Geometric means of the sulfate data from both types of mines were about 1,000 mg/L. However, waters that had reacted with mine tailings averaged over 18,000 mg/L at silica-carbonate mines and over 1,900 at hot-spring mines.

Chloride: Ashley et al. (2002) reported chloride levels in waters proximal to an abandoned placer gold mine in Shasta County, CA. Chloride content of these waters ranged from 1.7 to 24 mg/L with one value of 13,000 mg/L. Values within a range of 1.5 to 5 mg/L are reported to be in the normal range of surface waters near the Pacific Ocean. The extremely high value of 13,000 in one pond was thought to be a result of the influence of geothermal fluids.

pH: Ashley et al. (2002) reported pH levels in waters proximal to an abandoned placer gold mine in Shasta County, CA. pH of these waters ranged from 5.6 to 8.2 SU. The pH of the waters of San Carlos Creek below the New Idria Mine was in the range of 6.2 to 9.0, with the exception of the mine drainage itself, which had pH between 3.3 and 4.6 (Ganguli et al., 2000). Gray et al. (2000) reported pH of waters below Hg mines in Alaska to be neutral to slightly alkaline (pH 6.8 to 8.4). Rytuba reports that in silica-carbonate type deposits adjacent to mafic vents where large amounts of pyrite have been introduced into the native rock, mine drainage can be extremely acidic with pH as low as 2.4. Waters which drain deposits with less pyrite are weakly to moderately acidic and have pH ranging from 3.1 to 6.9.

Conductivity: Ashley et al. (2002) reported conductivity levels in waters proximal to an abandoned placer gold mine in Shasta County, CA. Conductivity of these waters ranged from 69.2 to 1320 uS/cm with one value of 36,300 uS/cm. As with chlorides, the extremely high value of 36,300 was thought to be a result of the influence of geothermal fluids. Gray (2003) reported conductivities associated with Hg mine drainages in Humboldt Co., NV between 350 to 510 uS/cm. Conductance in the Humboldt River downstream of the mine drainages was higher, ranging from 400 to 1000 uS/cm. Below the New Idria Mine, Ganguli et al. (2000) reported conductivity ranging from 3,180 to 5,240 uS/cm. Conductivities above the mine were much lower ranging from 800 to 1,120 uS/cm, respectively).

4.3.3 Mercury-Containing Sediments

Other sections of this report have primarily been concerned with oxic waters as this is the environment to which most wastewater treatment plants discharge. The enhancing and mitigating factors related to the bioavailability of Hg have primarily been identified and discussed in this context. However, when discussing the bioavailability of Hg in sediments, a slightly different set of factors must be considered. This is because methylation in sediments is primarily driven by sulfate reducing activity of sulfate-reducing bacteria (SRB) under low redox conditions. Controls on methylation have been reviewed by several authors including Gilmour and Henry (1991) and more recently Ullrich et al. (2001). Factors that appear to have the greatest effect of methylation in such environments are:

- Redox
- Partitioning of Hg^{II} to sediments/DOC
- Sulfate/sulfide levels
- Salinity

- Organic carbon/acid volatile sulfide content of sediments
- pH

Redox: Low redox appears to be essential for the methylation of Hg^{II} as the primary mechanism in natural waters appears to be sulfate reduction by sulfate reducing bacteria (SRB). These bacteria use sulfate as an electron acceptor in the process of degrading organic matter in sediments. The end product of sulfate reduction is sulfide. The primary factors controlling the metabolic activity of SRB are temperature, dissolved oxygen, organic carbon supply and sulfate supply. For instance, Compeau and Bartha (1984) showed that methylation of Hg was favored at redox of -220 mV versus +110 mV.

Partitioning of Hg^{II} to Sediments/DOC: Partitioning of Hg^{II} to sediments is important as is it thought that only dissolved Hg species are available to be methylated and therefore bioaccumulated. In filtered pore waters of Long Island Sound, Hammerschmidt and Fitzgerald (2004) found $\log K_D$ for Hg^{II} species to range from 3.24 to 4.76. $\log K_D$ for MeHg ranged from 1.78 to 3.01. MeHg production rates were also highest when AVS was low, and declined rapidly to an asymptotic rate above AVS levels of about 50 μM/kg. K_Ds were also shown to be affected by LOI and AVS. K_Ds for both MeHg and Hg^{II} were positively correlated with LOI and became asymptotic to constant values at levels of AVS above about 50 uM/kg. MeHg production levels were shown to decrease with increasing K_D of Hg^{II}. MeHg was found to be highly correlated with Hg^{II} in sediments.

In addition to being present in the dissolved phase, other geochemical considerations are likely to impact Hg^{II} methylation. It appears that the neutrally charged Hg^{II} species are more likely to be involved in methylation than charged species (Benoit et al., 1999; Drott et al., 2007). For instance, Barkay et al. (1997) noted that neutrally-charged Hg species (e.g. $HgCl_2$) were more bioavailable than negatively charged species. Increasing concentrations of DOC (i.e. increased binding to DOC) were shown to result in a decline in bioavailability of Hg^{II} species as well.

Sulfate/sulfide Levels: The presence of sulfate is obviously essential for SRBs to do their work. Several studies have shown that in low sulfate systems the addition of sulfate will stimulate the activity of SRB (Rudd et al., 1986; Jeremiason et al., 2006). Pak and Bartha (1998) found that methylation rates in the sediments of three oligotrophic lakes were positively correlated with dissolved sulfate levels in sediments. As previously pointed out, sulfide is the end product of sulfate reduction and when sulfide levels become too high, it can have an inhibitory effect. For example, Benoit et al. (1999) studied the influence of sulfide concentration of the methylation of Hg in sediments of the Patuxent River, MD and the Florida Everglades. Their conclusion was that pore water sulfide levels that are too high inhibit the formation of MeHg as it limits the amount of HgS^0 in solution, the neutrally-charged Hg-S species assumed to be most likely to pass through biological membranes and be biomethylated. Levels of HgS^0 decreased over four orders of magnitude as dissolved sulfide increased from 10^{-7} to 10^{-3} M. Lambertsson and Nilsson (2006) found that increasing sulfide concentrations in the pore waters of Bothnian Sea sediments to 5 μM (0.48 mg/L) decreased methylation by 20% relative to a control and increasing sulfide concentration to 350 μM (33.6 mg/L) decreased methylation activity by 90%.

Salinity: The methylating activity of marine and estuarine sediments is usually lower than freshwater sediments, partly due to salinity effects. The negative effect appears to be a result of the formation of charged sulfide and chloride complexes such as $HgCl_4^{2-}$. Thus estuarine fish tend to have lower MeHg in their tissue than comparable freshwater species (Gilmour and

Riedel, 2000). Several studies have noted the effect of salinity on methylation e.g. Compeau and Bartha (1984) who noted that methylation was favored under lower (0.4%) than higher (2.5%) salinities.

Organic Carbon/Acid Volatile Sulfide Content of Sediments: It is well-recognized that the bioavailability and toxicity of metals is a function of their relative association with the solid phase in sediments. One way to measure the availability of metals to biota is through the use of Simultaneously Extracted Metals/Acid Volatile Solids (SEM/AVS). This technique dissolves the acid volatile sulfides (i.e. sulfides or iron and manganese) and measures the metals released during the extraction. This technique has been used to characterize sediment toxicity in a number of studies (Lee et al. 2000; Patton and Crecelius, 2001; Helyer, 2000, Olinger, undated). Although it has been utilized less to assess the availability of Hg for bioaccumulation, the same principles apply. To the extent that metals concentrations exceed AVS, metals in sediments are judged to be less available to exert toxic effects, or, in the case of Hg, to be transformed to MeHg and/or bioaccumulated. It has been noted that when the organic carbon in sediments is accounted for (by dividing the (ΣSEM-AVS) by the fraction of organic carbon, f_{OC}), a significant improvement in prediction of toxicity results (McGrath et al., 2002). This measure of bioavailability thus takes into account some of the major factors influencing the concentration of T-Hg and MeHg in sediment pore waters including sulfides and organic carbon.

There is conflicting information on the effect of organic matter in sediments on methylation. For instance, Hammerschmidt and Fitzgerald (2004) found that while the concentration of T-Hg was positively correlated with the LOI of sediments, MeHg production rates were negatively correlated with LOI. On the other hand, Heim et al. (2007) found that MeHg in Sacramento/San Joaquin Delta sediments was positively correlated with LOI (R2 = 0.32). In pore waters, increased levels of DOC should result in lower Hg methylation rates due to the binding of Hg^{II} by DOC. Mason and Lawrence (1999) found that sediment organic content was a better predictor of MeHg in sediments than other factors including AVS. Lambertsson and Nilsson (2006) reported that OM was the most significant factor affecting factor affecting net methylation over a range on OM in sediments from 0.8 to 10.8%.

pH: The literature is also somewhat conflicting with regard to the effect of pH on methylation. As reported in Gilmour and Henry (1991) acidification of lakes has been observed to lead to higher methylation rates in water and sediment. However, studies in which sediments have been acidified have shown the opposite effect. For instance, Steffan et al. (1988) studied the effect of pH on methylation of Hg^{II} in sediments and found that acidification of sediments radically decreased methylation. Acidification of pH 6.1 sediments to 4.5 using sulfuric or hydrochloric acid decreased methylation by 65%. On the other hand, Kelly et al. (2003) found that decreasing pH and increasing DOC levels stimulated uptake of Hg^{II} by a freshwater bacterium.

4.3.4 Urban Runoff

Several of the parameters in this section were reported by Pitt et al. (2004) in the National Stormwater Quality Database (NSQD, Version 1.1). The database contains monitoring data collected over a 10-year period from more than 200 municipalities throughout the U.S. characterizing the quality of stormwater runoff. As of mid-summer of 2003, the database contained 3,770 separate events from 66 agencies in 17 states. Figure 4-5 taken from the USDA (1997) provides an overview of characteristic concentrations of suspended solids in streams draining predominately agricultural, urban, forest, and range watersheds.

Average concentration

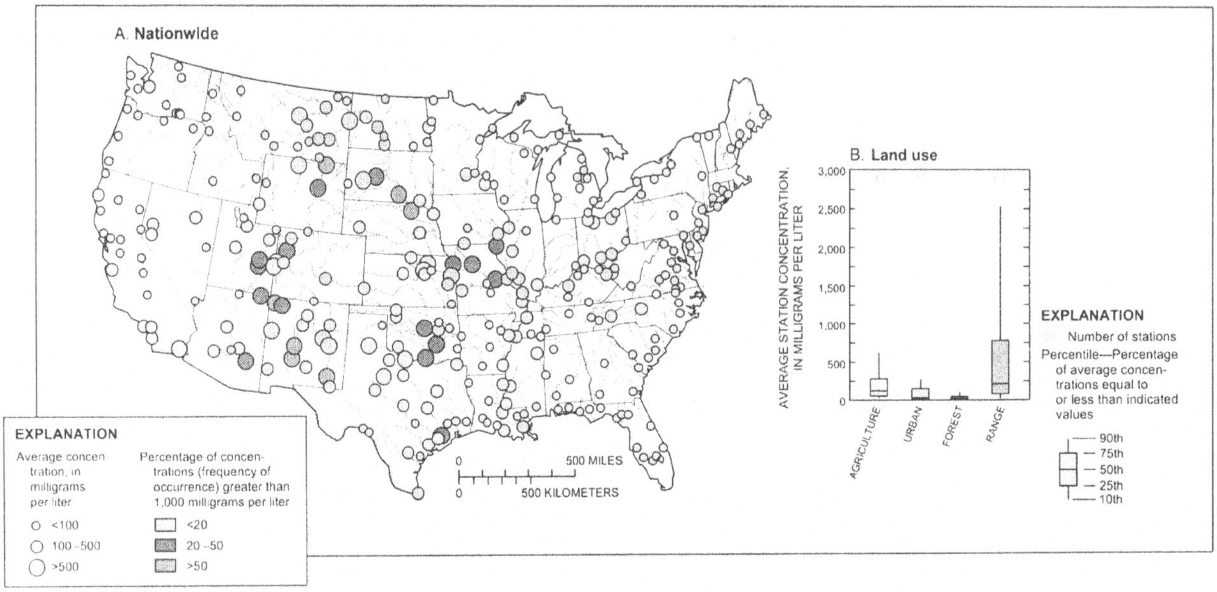

Figure 4-5. Overview of Suspended Sediment Concentrations in Streams Nationwide. Source: USDA 1997.

TSS/VSS: Pitt et al. (2004) report a median level of TSS in stormwater runoff of 58 mg/L. Gromaire-Mertz et al. (1999) report an event mean concentration (EMC) of 92.5 mg/L TSS from street runoff in the "Les Marais" catchment in central Paris, France. The median volatile suspended solids (VSS) concentration in street runoff was 550 mg/g (thus about 55% of the solids were volatile; that is, consisted of organic carbon). McKee et al. (2005) reported TSS flow- weighted means for water years 2003 and 2004 in the Guadalupe River were 204 and 191 mg/L respectively[16]. Babiarz et al. (1998) measured TSS in a Wisconsin river draining a predominately urban watershed to average 13.9 mg/L over an annual cycle. Measured values ranged from 3.2 to 73.7 mg/L. Median and 75th percentile TSS concentrations reported in CEPA (2005) about 80 mg/L and 120 mg/L during wet weather flows. These TSS values are similar to those reported for urban sites by Kinnetic Laboratories, Inc. (2002).

Mason and Sullivan (1998) report logK_D values for T-Hg and MeHg in two branches of the Anacostia River in urban Washington, D.C. Mean logK_D value (average of all values for both branches) was 4.15 ± 1.01 (n=24). For MeHg, the mean logK_D was 4.23 ± 0.73. While the K_D for MeHg is normally less than that of T-Hg, in this study, the K_D for MeHg is slightly higher.

DOC/POC/TOC: Pitt et al. (2004) report a median level of COD in stormwater runoff of 53 mg/L. Using the stoichiometric conversion discussed above the estimated value for TOC in

[16] Note that the constituents reported by McKee et al. (2005) were only measured during high flow events and are not necessarily comparable to data from other studies cited here where concentrations may have been measured over a wider range of flow conditions. For instance, T-Hg measurements made during the same period in 2003 yielded a range of concentrations from 180 to 18,670 ng/L with a flow-weighted mean of 2,190 ng/L and in 2004 ranged from 10 to 1,420 ng/L with a flow-weighted mean of 330 ng/L. These concentrations are very high when compared to other urban runoff values and are more similar to those reported for mine drainages, as pointed out by the authors.

urban stormwater would be 19.9 mg/L. The EMC for COD reported by Gromaire-Mertz et al. (1999) was 131 mg/L suggesting a TOC level of 49 mg/L. A median value of 66% of the COD was associated with particles in street runoff. Novotny and Chesters (1981) report the COD content of urban stormwater to range between 20 and 600 mg/L. Using the conversion discussed above to TOC, the TOC range would be 7.5 to 225 mg/L. McKee et al. (2005) report DOC ranges in the Guadalupe River, CA of 0.9 to 18.3 mg/L with flow-weighted means of 5.3 and 5.9 mg/L for water years 2003 and 2004. POC concentrations ranged from 0.1 to 4.3 mg/L with flow-weighted means of 2.3 and 1.0 mg/L for water years 2003 and 2004.

Babiarz et al. (1998) measured DOC in waters draining a predominately urban watershed and found the mean over an annual cycle to be 4.84 mg/L.

Sulfate: Soller et al. (2005) showed that median sulfate concentrations from runoff samples collected in the Guadalupe River watershed were about 10 mg/L from agricultural land-use sites (ranging from about 5 to 25 mg/L) whereas runoff from commercial/residential or mixed land use sites had concentrations of only about 1 mg/L. Gordon et al. (2005) reported sulfate levels in runoff from two urban drainages in Commerce City, CO. Data were collected between 1988 and 2004. Median sulfate concentrations were 32 mg/L and 21 mg/L respectively.

Polkowska et al. (2001) reported sulfate levels in roof and street runoff around the world including: Urungi, China, 3.37 to 9.79 mg/L; Lyon, France, 23 mg/L; M1 highway, United Kingdom, 60.4 to 247.9 mg/L; Minneapolis, MN, 5 to 650 mg/L, and Gdansk, Poland, 1.9 to 200.6 mg/L, mean of 44 samples 13.9 mg/L).

Chloride: Median chloride concentrations reported by Gordon et al. (2005) were 17.5 and 26 mg/L respectively in two urban Colorado drainages. Polkowska et al. (2001) reported chloride levels in roof and street runoff around the world including: Urungi, China, 0.54 to 1.43 mg/L; Lyon, France, 23 mg/L; M1 highway, United Kingdom, 65.1 to 229.1 mg/L; Minneapolis, MN, 0.96 to 46,000 mg/L, and Gdansk, Poland, 1.4 to 1409.5 (mean of 44 samples was 151 mg/L).

Although chloride levels would typically not be high enough in freshwater systems to affect the complexation of Hg, they may be elevated in runoff when used as deicing salts in cold climates. Specific conductance of highway runoff samples in such cases were reported by Granato and Smith (1999) to range between 37 and 51,500 uS/cm, the latter being similar to the conductivity of seawater. Log mean (2.35) and standard deviation (0.86) was reported for the chloride data. Given these statistics, the 95 percentile chloride concentration from this data set would be 5,300 mg/L (comparatively the Cl content of seawater is about 19,300 mg/L). These measurements were made off State Route 25 in southeastern Massachusetts from August 1988 through March 1995 and consisted on some 233 samples.

pH: Pitt et al. (2004) report a median level of pH in stormwater runoff of 7.5 SU. Gordon reported median pH values were 8.46 and 7.32 respectively in two urban Colorado drainages.

Conductivity: Pitt et al. (2004) report a median level of conductivity in stormwater runoff of 121 uS/cm. Gordon reported median conductivity values were 274 and 359 uS/cm, respectively, in two urban Colorado drainages. Babiarz et al. (1998) measured conductivity in an urban stream over an annual cycle. Values ranged between 483 and 1550 uS/cm with an average of 1004 uS/cm.

4.3.5 Non-Urban Runoff

TSS/VSS: Grigal (2002) reported values of $\log K_D$ for relationships between particle-bound and dissolved Hg, either reported or calculated from data in the literature, as ranging between 4.4 and 5.7 with a mean of 5.1 ± 0.4. According to a review by McKee et al., 2004, $\log K_D$s for freshwaters typically range between 3.7 and 6.6 and $\log K_D$ has been calculated for the Guadalupe River of 5.5 to 5.7. In relatively undisturbed watersheds, suspended sediment concentrations will be lower and a higher percentage of Hg will be in the dissolved phase. In more disturbed watersheds with higher TSS concentrations, the particulate phase is likely to dominate. Conaway et al. (2003) report $\log K_D$ values for T-Hg in suspended sediments in San Francisco Bay ranging from 3.4 to 6.7 with an average of 5.6 ± 0.65 (n=126). TSS ranged from 2 to 370 mg/L, with about 65% of the values falling between 10 and 100 mg/L.

TSS was measured by Bonzongo and Lyons (2004) in the Mobile River system, AL and ranged between 0 and 70 mg/L. Unfortunately, concentration on the solids was not measured. However, from the relationship between T-Hg (in ng/L) and TSS (in mg/L) in the water column (which had a slope of 0.0645), an average concentration on solids of 64.4 ng/g can be inferred. Using this value with the median dissolved T-Hg yields a $\log K_D$ of 5.2 L/kg. This value supports those reported above.

Balogh et al. (2003) report T-Hg and TSS levels in the Minnesota, Blue Earth and Le Sueur River in southern MN all of which drain predominantly cultivated areas. Slopes of the T-Hg/TSS regression lines were reported to two locations on the Minnesota River as being 52 ng/g and 46 ng/g, respectively. Slopes for the Blue Earth and Le Sueur Rivers were reported to be 51 ng/g and 48 ng/g, respectively. Dissolved T-Hg concentration was indicated to be very near 1 ng/L in all these systems. Given this, and assuming a characteristic T-Hg in suspended sediment concentration of 50 ng/g, a $\log K_D$ of 4.7 can be calculated. The resulting value is in the ranges reported above. The slopes of the regressions of T-Hg and TSS for these rivers are reported to be very similar to the T-Hg concentrations in the silt and clay fraction of soils across the north-central U.S. (52 ng/g).

Babiarz et al. (1998) observed TSS concentrations in streams draining non-urban watersheds ranging from 0.91 mg/L in a forested watershed to 1005 mg/L in an agricultural watershed. Mean TSS values were highest from the predominately agricultural watershed (125 mg/L) and lowest from the predominately forested watershed (1.94 mg/L). Mean TSS from other non-urban watersheds ranged from 4.8 to 19.1 mg/L. As described earlier in section 4.2.5, $\log K_D$s were found to range between 2.8 and 5.5 for T-Hg and 2.9 and 5.8 for MeHg and were found to be comparable with other studies. The effect of increasing TSS concentration on the magnitude of the partition coefficient was noted, with slopes of the K_D versus TSS relationship being statistically zero to negative. Over all sites, increasing TSS from near zero to near 100 mg/L affected K_D by one to two orders of magnitude.

DOC/POC/TOC: Kolka et al. (2001) showed that mean monthly concentrations of T-Hg in runoff was well-correlated with POC in upland surface runoff, upland interflow, bog flow and streamflow from a northern Minnesota forested watershed. POC levels ranged up to 20 mg/L but were typically 10 mg/L or less. Pettersson et al. (1995) reported DOC levels in peatland watershed in northern Sweden ranged from 10 to 40 mg/L. Grigal (2002) reports that T-Hg concentrations are well-correlated with DOC and slopes of relationships from North America have slopes of about 0.2 ng-T-Hg / mg-DOC. The relationship between T-Hg and DOC degrades during these high flow events where particle bound Hg dominates the flux. Particle-bound Hg is associated with the organic carbon content of the sediments, suggesting that the Hg is associated with POC. It has been noted by Foster (1980) that organic matter is usually enriched in eroded sediment over watershed soils by a ratio ranging from 1.2 to 4.4. Therefore, if OM content of watershed surficial soils are available, then estimates can be made of the OM content of eroded sediments.

FTN Associates (2002) retrieved STORET data for TOC in six watersheds in central Arkansas in the course of preparing Hg TMDLs for EPA Region 6. TOC levels ranged from < 5 to > 10 mg/L. Mason and Sveinsdottir (2003) measured DOC and POC in 20 Maryland reservoirs. Values for TOC averaged 3.66 ± 1.96 (n=37) and were compared to values measured in Wisconsin lakes averaging 7 mg/L. Average POC concentration was 0.12 ± 0.07 (n=38). Canham et al. (2004) reported DOC in 355 Adirondack headwaters lakes. Values ranged from near 0 to 26 mg/L but the majority of values were under 15 mg/L. Gergel et al. 1999 report DOC concentrations for Wisconsin lakes and rivers. DOC from 44 drainage lakes averaged 6.0 ± 3.0 mg/L, 66 seepage lakes averaged 5.5 ± 3.4 mg/L and nine spring-fed lakes averaged 7.2 ± 3.6 mg/L. Sampling at 24 river sites showed an average DOC of 10 ± 6.9 mg/L. Nine boreal lakes studies by Hudson et al. (2003) had DOC concentrations ranging from 1.8 to 5.1 mg/L. Balogh et al. (2004) reported DOC concentrations between 5 and 25 mg/L from two streams draining areas comprised of 40 to 50% agricultural land. Conaway et al. (2003) reported that DOC values ranged over two orders of magnitude from 59 uM to 890 uM (0.71 to 10.7 mg/L) in the San Francisco Bay estuary, although most values were in the 100 to 300 uM (1.2 to 3.6 mg/L) range.

Babiarz et al. (1998) observed a positive relationship between DOC and filtered T-Hg in samples from streams draining six watersheds of varying land use in WI. DOC concentrations ranged from 1.48 mg/L in a stream sample in a predominately agricultural watershed to 41.1 mg/L in drainage waters from a predominately wetland watershed. Lowest mean DOC level were associated with the agricultural watershed (2.51 mg/L) and the highest mean level was associated with the wetland watershed (24.7 mg/L).

Karlsson and Skyllberg (2003) measured binding constants of MeHg to NOM in soils by displacing native MeHg with various halides. Results show log binding constants on the order of 15.6 to 17.2 with higher binding constants calculated at the lowest pH (pH = 2) and the lowest calculated at the highest pH (pH = 5). The authors also report binding constants measured by others ranging from $\log K_D = 12.15$ to 14.48 for humic and fulvic acids, $\log K_D = 10.4$ to 14.8 for humic acids, and $\log K_D = 16.7$ to 17.1 for binding sites in an organic soil.

Sulfate: FTN Associates (2002) retrieved STORET data for sulfates in six watersheds in central Arkansas in the course of preparing Hg TMDLs for EPA Region 6. These watersheds were primarily forested > 70% with some limited agriculture and small communities. Sulfate levels were predominantly in the 5 to 35 mg/L range. In the Ouachita River Basin, impacted by the Hg mining area in the Ouachita mountains of Arkansas, sulfate levels appeared to be about the same, to slightly higher, with most in the 5 to 25mg/L range, but with some exceeding 35 mg/L. Bayou Cortableau, a predominately agricultural watershed (64% of land use) with a significant percentage of wetlands (32%) in the Vermilion-Teche Basin in Louisiana were reported to have sulfate levels ranging from 4.5 to 55.7 mg/L over the measurement period from June to December 1998 (U.S. EPA, 2000a). Lillie et al. (1983) show that Wisconsin lakes range from < 10 in the north to > 40 mg/L sulfate in the southeastern part of the state. Mason and Sveinsdottir (2003) measured sulfate in 20 Maryland reservoirs. Values for sulfate averaged 9.49 ± 2.80 (n=38) and were said to be comparable to measurements in other freshwater systems. Balogh et al. (2004) reported sulfate levels between 1 and 25 mg/L for two small watersheds in northcentral Minnesota with roughly 40-50% agricultural land use. Sulfate levels were lower during stormflows when it was speculated that wetlands were primary source areas for runoff. According to the World Health organization (WHO, 2004) typical sulfate concentrations in fresh water are in the vicinity of 20 mg/L and range from near zero to 630 mg/L in rivers and from 2 to 250 mg/L in lakes.

Bonzongo and Lyons (2004) report that sulfate had an effect on MeHg over a narrow range for inland waters of the Mobile River system in Alabama. The entire range of sulfate was near zero to nearly 600 mg/L, with all values > 100 mg/L occurring in the Bay. In the Bay, MeHg concentrations were insensitive to sulfate levels. In inland waters, however, %MeHg was related to sulfate as evidenced by the regression:

$$\%MeHg = 0.9245(SO4) - 6.886 \qquad 4.3.5\text{-}1$$

which had an R^2 of 0.825.

Chloride: Mason and Sveinsdottir (2003) measured chlorides in 20 Maryland reservoirs. Values averaged 33.8 ± 28.6 (n=38), were said to vary widely across reservoirs, and to be higher than observations in Wisconsin lakes (0.11 to 4.0 mg/L). A possible explanation for high Cl levels in several reservoirs was snowmelt carrying road deicing salts. According to the World Health Organization (WHO, 1996) chlorides in unpolluted waters are often below 10 mg/L and sometimes below 1 mg/L. Mean chloride levels in several rivers of the United Kingdom ranged from 11 to 42 mg/L between 1974 and 1981. Chloride levels in arid areas of the US west may have elevated concentrations of chlorides due to accumulation of salts in soils. Mashburn and Sughru (2004) measured chlorides at 11 "chloride impacted" sites in Oklahoma and found median levels ranging from 196 to 2,639 mg/L.

pH: FTN Associates (2002) retrieved STORET data for pH in six watersheds in central Arkansas in the course of preparing Hg TMDLs for EPA Region 6. pHs were predominantly in the 5.5 to 6.5 range. Balogh et al. (2004) reported sulfate pH between 7.2 to 7.8 for two small watersheds in northcentral Minnesota with roughly 40 to 50% agricultural land use. pH was lower (7.2 to 7.4) during stormflows when it was speculated that wetlands were primary source areas for runoff.

Bonzongo and Lyons (2004) report that pH had an effect of MeHg over a very narrow range (6.8 to 7.1) in the Mobile River system. Within this range a regression was developed for %MeHg as a function of pH, being:

$$\%MeHg = 13.18 - 1.85pH \qquad 4.3.5\text{-}2$$

with a R^2 of 0.96. Outside of this range, MeHg was insensitive to changes in pH.

Conductivity: According to Boyd (1979) natural waters have conductivity of 20 to 1500 uS/cm. Babiarz et al. (1998) measured conductivity in waters of six non-urban watersheds in Wisconsin.

Mean conductivities ranged from 124 uS/cm (forested watershed) to 678 uS/cm (agricultural watershed).

4.3.6 Wastewater Treatment Plants

Tables 4-9 and 4-10 show the effluent water quality associated with typical secondary and advanced wastewater treatment plants. While this publication is somewhat dated, it provides a reasonable snapshot for the parameters of interest. Recent advances in wastewater treatment have focused on nutrient removal and are not expected to have impacted the parameters of interest to this report to any great extent.

TSS/VSS: Suspended solids from the secondary treatment plants in Table 4-9 range from 13 to 26 mg/L and from 1 to 5 mg/L in the advanced treatment effluents shown in Table 4-10. TSS levels reported by Novotny and Chesters (1981) for treated effluent range from 10 to 30 mg/L and are typical of secondary treatment levels.

DOC/POC/TOC: Although the TOC or DOC of these effluents is not reported, COD from one secondary plant in Table 4-9 has a reported value of 27 mg/L. For comparative purposes, Novotny and Chesters (1981) report the COD of municipal wastewater effluent to range from 25 to 80 mg/L. Using the conversion discussed above to TOC, the TOC range would be 9 to 30 mg/L. Eckenfelder (1970) reports the ratio of 5 day BOD to TOC to be about 1.87. Therefore, multiplication of the BOD_5 values from Tables 4-9 and 4-10 would also give representative values of TOC. Based on the BOD in effluent from secondary treatment plants, one would expect TOC to be in the range of 4 to 11 mg/L. Using this same approach, TOC values from the advanced WWTPs would be expected to be in the range of 1 to 5 mg/L.

Sulfate: Sulfate is reported to range from 66 to 192 mg/L in the secondary plants and 123 to 212 mg/L in the advanced WWTPs above. Therefore, it does not appear that advanced treatment affects sulfate levels. The very recent measurements of sulfate in the San Jose WWTP (see Chapter 3.0) of 100 to 120 mg/L are reasonably similar.

Chloride: Chloride levels are reported to range between 81 and 245 mg/L in the secondary plants and 105 to 184 at the advanced treatment plants above. Again, the very recent measurements in the San Jose WWTP show similar levels (150 to 200 mg/L; see Chapter 3.0).

pH: pH in the one secondary treatment plant reported above was 7.6 and the advanced treatments plants were reported to have values of 6.8 and 7.1.

Conductivity: Conductivity reported for the secondary plant above is 1,390 uS/cm. For the advanced WWTPs the range is from 1020 to 1440 uS/cm. Since natural waters typically have conductivities ranging between 20 to 1500 uS/cm, conductivities for WWTPs would appear to be at the upper end of the natural water range.

Table 4-9. Quality of Secondary Effluent at Selected Wastewater Treatment Plants in California.

Quality parameter (mg/L except as otherwise indicated)	Plant location			
	Trickling filters		Activated sludge	
	Chino Basin MWD (No. 1)	Chino Basin MWD (No. 2)	Santa Rosa Laguna	Montecito Sanitary District
Biochemical oxygen demand, BOD_5	21	8	-	11
Chemical oxygen demand	-	-	27	-
Suspended solids	18	26	-	13
Total nitrogen	-	-	-	-
NH_3-N	25	11	10	1.4
NO_3-N	0.7	19	8	5
Org-N	-	-	1.7	-
Total phosphorus	-	-	12.5	-
Ortho-P	-	-	3.4	-
pH (unit)	-	-	-	7.6
Cations:				
Ca	43	55	41	82
Mg	12	18	18	33
Na	83	102	94	-
K	17	20	11	-
Anions:				
HCO_3	293	192	165	-
SO_4	85	143	66	192
Cl	81	90	121	245
Electrical conductivity dS/m	-	-	-	1.39
Total dissolved solids	476	591	484	940
Sodium adsorption ratio	2.9	3.1	3.9	3.7
Boron (B)	0.7	0.6	0.6	0.7
Alkalinity ($CaCO_3$)	-	-	-	226
Total Hardness ($CaCO_3$)	156	200	175	265

Source: Asano and Tchobanoglous (1987)

Table 4-10. Effluent Quality Data from Selected Advanced Wastewater Treatment Plants in California[1].

Quality parameter (mg/L except as otherwise indicated)	Plant location					
	Long Beach	Los Coyotes	Pomona	Dublin San Ramon	City of Livermore	Simi Valley CSD
Biochemical oxygen demand, BOD_5	5	9	4	2	3	4
Suspended solids	-	5	-	1	-	-
Total nitrogen	-	-	-	-	-	19
NH_3-N	3.3	13.6	11.4	0.1	1.0	16.6
NO_3-N	15.4	1.1	3	19.0	21.3	0.4
Org-N	2.2	2.5	1.3	0.2	2.6	2.3
Total phosphorus	-	-	-	-	-	-
Ortho-P	30.8	23.9	21.7	28.5	16.5	-
pH (unit)	-	-	-	6.8	7.1	-
Oil and grease	-	-	-	-	-	3.1
Total coliform bacteria, MPN/100 ml	-	-	-	2	4	-
Cations:						
Ca	54	65	58	-	-	-
Mg	17	18	14	-	-	-
Na	186	177	109	168	178	-
K	16	18	12	-	-	-
Anions:						
SO_4	212	181	123	-	-	202
Cl	155	184	105	147	178	110
Electrical conductivity, dS/m	1.35	1.44	1.02	1.27	1.25	-
Total dissolved solids	867	827	570	-	-	585
Soluble sodium, %	63.2	59.2	51.7	-	-	-
Sodium adsorption ratio	5.53	4.94	3.37	4.6	5.7	-
Boron (B)	0.95	0.95	0.66	-	1.33	0.6
Alkalinity ($CaCO_3$)	-	256	197	150	-	-
Total Hardness ($CaCO_3$)	212	242	206	254	184	-

[1]Advanced wastewater treatment in these plants follows high rate secondary treatment and includes addition of chemical coagulants (alum + polymer) as necessary followed by filtration through sand or activated carbon granular medium filters.

Source: Asano and Tchobanoglous (1987)

CHAPTER 5.0

METHODS TO ESTIMATE BIOAVAILABILITY OF MERCURY IN EFFLUENTS AND RECEIVING WATERS

5.1 Summary

Definitions of bioavailable mercury, derived earlier, are used to construct a framework and general approach to assessing the relative bioavailability of Hg in WWTP effluents and other sources. The focus is on the bioavailability of Hg in the source waters and this chapter does not deal in any detail with the impact of mixing with receiving waters on the bioavailability, as this will be covered in detail in future work. The approach taken is also to define and estimate the fraction that is bioavailable rather than the total amount of bioavailable Hg, as this makes comparison between systems easier and does not rely on factors such as large differences or variability in total Hg concentrations or loading rates.

The approach taken for the methods outlined in this chapter is based on the assumptions that all the methylmercury (MeHg) in the water is bioavailable and that only a fraction of the dissolved inorganic Hg is bioavailable, while particulate inorganic Hg is not. Furthermore, it is discussed and concluded that the dissolved inorganic Hg bound to dissolved natural organic matter may be considered essentially non-bioavailable. Therefore, the fraction of the total Hg that is bioavailable, denoted as fB-Hg, is equivalent to the fraction as MeHg plus the dissolved reactive Hg (R-Hg) fraction.

The chapter provides alternative ways of determining each of the method parameters or surrogates for these parameters, and discusses the types of estimations that can be made, and their certainty, which is a function of the amount and extent of the information available. Estimates of fB-Hg can be made if only the total Hg concentration (T-Hg) and the suspended solids concentration (TSS) is known, but such estimates will result in a range of values, and the validity of the results should be checked using sensitivity analysis. Such an analysis would provide an indication of parameters that may need to be measured to improve the estimate of fB-Hg. Clearly, more detailed information will allow for a more rigorous estimate of fB-Hg. There are other possible approaches to estimating the fB-Hg but given the focus of this work, it is not relevant to detail the specifics of other approaches. The methodology outlined here is applied to several specific datasets to provide a “proof-of-concept” of the research team’s particular approach.

5.2 Introduction

Before proceeding with the estimation of bioavailability, it is pertinent to recall the definitions that were outlined in Chapter 2.0 and which provide the guiding principle for the discussion below. Bioavailable mercury is defined in Chapter 2.0 as:

...those forms of mercury, both inorganic and organic, that can be readily transported across biological membranes, either by active or passive processes and subsequently accumulated, and/or biotransformed to a more toxic and bioaccumulative species.

The definition encompasses all forms of inorganic Hg and organic Hg compounds. For the purposes of this study, the only form of organic Hg that will be considered is methylmercury (MeHg) and, additionally, it is assumed that all MeHg in a particular waterbody is bioavailable, even if associated with particles. In most waters, given the low K_D for MeHg relative to inorganic Hg – typically, at least an order of magnitude lower – the majority of the MeHg is in solution. However, in instances of very high particulate loading, the approach may lead to an overestimate of bioavailability as some of the particulate associated MeHg may be associated with particles and not be bioavailable. Clearly, however, the particulate MeHg fraction includes both MeHg associated with inorganic particles and MeHg associated with biotic particles, such as phytoplankton and bacteria, or biotic-derived detritus, and the latter represents a bioavailable fraction as it has already been assimilated. In most instances, because of the binding characteristics of MeHg, the amount of MeHg associated with inorganic material will be small. Also, in many waters the fraction of total Hg that is MeHg is small (<10%) and so this simplification of the definition will be valid in most instances. For high particulate scenarios, it would also be more valid to estimate particle bound MeHg using only on the fine fraction (e.g. < 63 μm) as most of the particle bound MeHg would reside in this more reactive fraction.

Given this definition, there are likely to be very few waters where there is no bioavailable Hg. Furthermore, the researchers make no distinction between bioavailable inorganic Hg and MeHg. In terms of bioavailability to the base of the food chain, or the incorporation of Hg into bacteria and Hg methylating organisms, this lack of distinction is valid, but it should be noted that in terms of bioaccumulation through the food chain, a larger fraction of the MeHg is bioaccumulated to the higher trophic levels than of inorganic Hg. However, there is the potential for bioavailable inorganic Hg to be methylated to MeHg and then bioaccumulated, and this is the principal reason for including the bioavailable inorganic Hg fraction in the estimate. It has been noted that in some ecosystems, Hg is methylated and demethylated rapidly with a lifetime of MeHg, on the order of days in some locations (sediment porewaters, for example) and therefore it is necessary to track the overall bioavailable fraction of Hg and MeHg and not just MeHg alone.

Note further, as discussed below, that a form of Hg not currently bioavailable can over time be transformed into a bioavailable form. Therefore it is necessary to stipulate that the term bioavailable Hg represents bioavailability of Hg and MeHg in its current form. In addition, therefore, it is necessary to provide a working definition of the term *potentially bioavailable Hg and MeHg*:

Potentially bioavailable mercury is any form of mercury that can be converted into bioavailable mercury in a scientifically reasonable time span.

As discussed in the previous chapters, the definition of a "reasonable time span" depends on many factors but for the purposes of defining bioavailability and estimating the bioavailable fraction, the time span is the time until the Hg from a particular source becomes well-mixed and indistinguishable from that Hg already in the receiving waters, which is considered to be days to weeks, based on the literature (e.g. Lamborg et al., 2004; Hsu-Kim and Sedlak, 12005; Miller, 2005; Hintelmann and Harris, 2004; see Chapter 3.0). Given that all MeHg is considered bioavailable, the definition of the potentially bioavailable fraction relates only to inorganic Hg.

Mostly, Hg^{II} bound to particles is only slowly released back into solution and, as this fraction is not considered bioavailable under the definitions above, the timescale of desorption helps determine the definition of the "reasonable timescale". The results of the study by Hintelman and Harris (2004) indicate that the half-life for desorption is on the order of a week, and thus, the timescale based on this criteria must be less than that value. Additionally, the timescale over which Hg can exchange between fractions in solution, and especially for its exchange reactions in the presence of NOM, is on the order of a day so that such re-equilibration, if it occurs, will happen within the timescale of the proposed definition. Given these considerations, a reasonable time span is hours to days, rather than weeks to months.

Note that there is insufficient data available to assess quantitatively the impact of mixing on relative bioavailability due to changes in speciation other than those that can be estimated using thermodynamic equilibrium calculations, but clearly this is an important consideration. It should be noted that while the rate of desorption of Hg from particles is slow, this is not the case for adsorption and thus it is possible to have adsorption occur upon mixing. In the context of the discussions below, it is assumed that mixing does not have any impact besides that associated with a conservative mixing of Hg fractions, and the resultant speciation changes that occur are based on chemical equilibrium fundamentals. This appears reasonable if the mixing does not dramatically alter the concentration on a per gram basis for the suspended matter as it is this Hg concentration, rather than the TSS concentration, that determines the dissolved Hg concentration (assuming equilibrium partitioning conditions are reached). In studies of the impact of sediment resuspension, Kim et al. (2004) found in studies of simulated tidal resuspension of particles that there was little change in the water column dissolved Hg concentration under widely varying TSS conditions. Therefore, as long as there was no substantial change in the "character" of the suspended material and in its Hg binding capacity upon mixing, there should be little change in the dissolved-particulate partitioning upon mixing.

5.3 An Approach to Estimating Bioavailability

As discussed above, not all the Hg in effluents or other source waters is in a bioavailable form and to assess the relative bioavailability of sources it is necessary to determine the bioavailable fraction. It should be pointed out that there are two ways of approaching such an estimation and discussion. The approach that will be taken here is an assessment of the relative bioavailability of different sources, which will be represented by the acronym fB-Hg, and such an approach does not take into account the actual total bioavailable Hg concentration (denoted as B-Hg), nor the potential loading from a particular source. Rather, fB-Hg is an assessment of how much, as a fraction of the total Hg, is bioavailable. Also, in making such estimates, the bioavailability estimate is based on the total Hg concentration in an unfiltered water sample (T-Hg) as this is the most widely measured and available parameter. Clearly, in terms of estimating relative contributions of various sources to receiving waters, it is necessary to take into account both the concentration of Hg in the various sources and their flow rates (F) (in units of volume/d). This will not be the approach taken in this chapter, even though it is the focus of TMDL allocations and other regulatory accounting. Given the definitions above, then[17]:

[17] In this chapter, variable names having a hyphen, such as T-Hg, when used in an equation, are underscored to designate the variable name T-Hg rather than a subtraction.

$\underline{B\text{-}Hg} = f\underline{B\text{-}Hg} * \underline{T\text{-}Hg}$ 5.3-1

and the flux of bioavailable Hg from a particular source would be:

Bioavailable Hg Flux = F * $\underline{B\text{-}Hg}$ (units of ng/d) 5.3-2

or

Bioavailable Hg Flux = F * $f\underline{B\text{-}Hg} * \underline{T\text{-}Hg}$ 5.3-3

Given these definitions it is clear that it is possible to have a high bioavailable Hg flux to a receiving water even if the fraction of the total Hg that is bioavailable (fB-Hg) is small under a situation where the value of either F or T-Hg was much higher for one source relative to the others.

There are many factors that go into the estimation of fB-Hg. However, from a review of the literature, as discussed in the earlier chapters, it is apparent that this value will vary from around 0.9 for waters where Hg is highly bioavailable (such as remote rain water) to values of <0.1 (for high particulate, high NOM, low methylation waters). However, for many waters, it appears that the value likely ranges between 0.1 and 0.4.

5.4 Bioavailability Calculations

5.4.1 The Bioavailability of Inorganic Particulate Hg

In terms of the definition of fB-Hg, particulate-associated inorganic Hg will be assumed to be unavailable on the timescales that are used to define both bioavailable Hg and potentially bioavailable Hg. This is a reasonable assumption as inorganic Hg has one of the highest binding constants (partition coefficients) for metals and is most likely associated in the solid phase with either inorganic sulfide ligands or thiol groups in particulate organic matter (POM). These associations are strong and dissociation of such complexes is kinetically slow. This result is consistent with the results of Hintelman and Harris (2004) that investigated the rate of adsorption and desorption of inorganic Hg and MeHg from sediment particles. They concluded that while adsorption was rapid and reached equilibrium within a day, desorption was not evident over a number of days. Kim et al. (submitted) modeled the adsorption and desorption of MeHg from particles and used a value equivalent to a desorption rate of ~4% d^{-1}. The rate of desorption of inorganic Hg is less as it is more strongly bound and thus, given the definition of bioavailability and potential bioavailability, the assumption that particulate Hg is not bioavailable appears valid. So, for a particular waterbody, the first task in estimating bioavailability will be an estimation of the dissolved fraction of mercury. Approaches to doing such estimates are discussed below. Bioavailable inorganic Hg is a fraction of the dissolved Hg pool and this fraction will depend on NOM and other factors that can alter the relative bioavailability. As noted above, given the definitions of bioavailability, the bioavailable fraction includes dissolved inorganic Hg species and dissolved or particulate bound MeHg.

5.4.2 Reactive Mercury and Methylmercury

If the MeHg concentration has been measured or can be estimated with sufficient accuracy under a specific set of conditions, then the difficulty in the estimation of bioavailability essentially is in the assessment of the bioavailable inorganic Hg fraction. As noted in earlier chapters, one operationally-defined measure of bioavailable inorganic Hg is the reactive Hg measurement (R-Hg) (e.g. Lamborg et al., 2003) and it is proposed that for the purposes of the estimation of fB-Hg, this is a good surrogate measurement for dissolved bioavailable inorganic Hg (IB-Hg), as discussed in Chapter 2.0. Note that the R-Hg measurement technique does not measure MeHg. Thus, if such measurements are available or can be estimated, then these should be used to estimate the fB-Hg.

Therefore, it is possible to estimate fB-Hg as:

$$\underline{\text{fB-Hg}} = (\text{MeHg} + \underline{\text{R-Hg}})/\underline{\text{T-Hg}} \qquad 5.4\text{-}1$$

where MeHg is total MeHg in the water column, and R-Hg is the reactive water column Hg, which is normally measured on a filtered water sample. An alternative formulation of Equation 5.4-1 is therefore also:

$$\underline{\text{fB-Hg}} = (\text{MeHg} + \underline{\text{IB-Hg}})/\underline{\text{T-Hg}} \qquad 5.4\text{-}2$$

If the %MeHg and %R-Hg is known or can be estimated from the literature, then fB-Hg can be estimated directly, as a fraction or as a percentage. While MeHg concentration is not always measured for effluents and other sources waters, there are values in the literature that can allow a reasonable estimation of the fraction of Hg as MeHg. Such an approach was used, for example, in constructing Table 4-1 in Chapter 4.0, where the %MeHg and the %R-Hg were listed for a variety of water types. As can been seen in that table, there is more information available about the MeHg fraction compared to the R-Hg fraction.

Given the data in the table, it is possible to obtain a range of estimates for fB-Hg for the various sources discussed in Chapter 4.0. The values of %MeHg in Table 5-1 are all based on literature values, and in all cases there are multiple sources of information, as noted in the discussions in Chapter 4.0. For R-Hg, there is some data in the literature for R-Hg in atmospheric deposition for a number of locations (Mason et al., 1992; 1997; Lamborg et al., 1995), one report of R-Hg for mining runoff (Gray et al., 2000) and Balcom et al. (2004) provide data from their study of rivers and effluents flowing into Long Island Sound for WWTP effluents and receiving waters. This last dataset was used to provide an estimate of R-Hg for non-urban runoff, and by extrapolation, for urban runoff. The estimated range of R-Hg in WWTPs was based on the results of work done in Phase II of this project.

Table 5-1. Estimated Fraction of Dissolved Mercury Considered Bioavailable, Using Ranges in the Estimated Fraction of the Dissolved Mercury that is Reactive, and Ranges in Methylmercury Concentration, According to Equation 5.4-1, for Different Source Water Types.

Water Type/Source	%MeHg	%R-Hg	Range in estimated bioavailable Hg fraction
Atmospheric Deposition	1-5	15-95	0.2 - > 0.9
Mining Runoff	0.1-1	1-5	< 0.1
Impacted Sediment Pore water	10-25	5-20	0.15 – 0.45
Unimpacted Sediment Pore water	10-25	10-20	0.2 – 0.45
Runoff - urban	1-2	5-10	< 0.1
Runoff - nonurban	5-15	15	0.2 – 0.3
Wastewater Treatment Plant Effluent	1-10	5-25	< 0.3

In the consideration of the potential of sediments as a source, it is likely that only porewater Hg is readily bioavailable and this fraction needs to be estimated. There is also some information available based on the "reactive Hg" measurements of sediment made by Marvin-DiPasquale and his colleagues (Marvin-DiPasquale, personal communication) and these data were used to estimate the %R-Hg for different pore waters. Based on the measurements, and the chemistry of porewater and data in the literature, it is estimated here that the %R-Hg would range from 5-20% for impacted sediments and 10-20% for unimpacted sediments. With these estimations, it is then possible to use the range of values to determine the range in fB-Hg for the various waters and sources using Equation 5.4-1, as shown in Table 5-1. An examination of the data in the table suggests that WWTP effluents have a relatively low fB-Hg, not as low as that of mining or urban runoff, but lower than that of atmospheric deposition or sediment pore water. Clearly, the bioavailable fraction is the highest for atmospheric deposition and this is the result of the high %R-Hg found in many rain waters, which do not have high %MeHg. In contrast, the intermediate value of fB-Hg for pore waters is driven by the high relative amount of MeHg in such waters.

It should be noted that there are two approaches for determining the fB-Hg of any particular water. The first approach is direct measurement. The second approach is that used in the construction of Table 5-1, i.e. use of values in the literature, or use of values that are deemed reasonable estimates based on the known chemistry of Hg and MeHg in environmental waters. While this approach cannot give a definitive answer, as illustrated by Table 5-1, it can provide a range of values that is, for many waters, relatively well-constrained. As noted above, in many, or even most instances in the literature, R-Hg is not measured and so, it is necessary to provide alternative ways to estimate fB-Hg, or to estimate the "non-bioavailable (non-reactive) dissolved inorganic Hg fraction" (INR-Hg), which would then allow an estimation of R-Hg as R-Hg ~ dissolved I-Hg less INR-Hg.

However, it is possible to make estimates of the INR-Hg or R-Hg (IB-Hg) fraction based on the NOM content, the dissolved sulfide content and other parameters. If INR-Hg can be estimated, then the IB-Hg fraction can be estimated if the dissolved inorganic Hg concentration (I-Hg) is known. This estimate requires a number of assumptions that are discussed further below.

5.4.3 Estimating Dissolved Bioavailable Inorganic Mercury

There are other methods and other data that could be used to estimate fB-Hg, especially if there is no data on R-Hg, or if an approach is needed that does not rely on the use of an operationally-defined measurement. Also, there will be instances where there is no information of MeHg, and thus there needs to be a mechanism to estimate bioavailability when this parameter has not been measured. Given the discussion above, the following definition is another valid approach:

$$f\underline{\text{B-Hg}} = [\text{MeHg} + (\underline{\text{T-Hg}} - \text{MeHg} - \underline{\text{IP-Hg}} - \underline{\text{INR-Hg}})]/\underline{\text{T-Hg}} \qquad 5.4.3\text{-}1$$

which simplifies to:

$$f\underline{\text{B-Hg}} = (\underline{\text{T-Hg}} - \underline{\text{IP-Hg}} - \underline{\text{INR-Hg}})/\underline{\text{T-Hg}} \qquad 5.4.3\text{-}2$$

or

$$\%(\underline{\text{B-Hg}}) = 100 - \%(\underline{\text{IP-Hg}}) - \%(\underline{\text{INR-Hg}}) \qquad 5.4.3\text{-}3$$

where IP-Hg is particle-bound inorganic Hg.

Both the non-reactive dissolved inorganic Hg fraction (INR-Hg) and the particulate inorganic Hg fraction (IP-Hg) need to be determined to use this approach. While the advantage of this calculation is that there is no need to measure MeHg concentration directly, there is also a need for caution as many of the parameters needed are not routinely measured. Having said this, a method that does not require MeHg determination is useful as the determination of MeHg is neither routine or trivial and is more costly than measuring total Hg concentration. The estimation method required in the absence of MeHg data has the potential for more error as there are more parameters that need to be measured or estimated in the calculation. Specifically, both IP-Hg and INR-Hg are not typically measured. Note that the total Hg measurement, of either particulate material or of filtered water includes the fraction as MeHg. For the particulate phase, however, as most suspended particulate matter (TSS) in effluent and natural waters has a low %MeHg, usually less than 10%, it is reasonable under such circumstances to substitute the total Hg suspended solid concentration in the calculations of the particulate inorganic Hg fraction (i.e. assume IP-Hg ~ P-Hg). Of course, if the %MeHg in the TSS were higher than 10%, then there would be substantial error in making this assumption. A more accurate estimate would require knowledge of the amount of MeHg in the solid phase so that this could be taken into account (i.e. IP-Hg = P-Hg – P-MeHg). The potential error in making an estimate without considering P-MeHg is shown in the example calculation at the end of this chapter.

Similarly, to estimate the INR-Hg fraction requires knowledge of how much inorganic Hg there is in solution, and the mechanisms to estimate this value have the potential to introduce error to the overall estimate. There are potential errors that could occur, for example, if the dissolved fraction is estimated as T-Hg – P-Hg as both measurements include the MeHg in these phases. If the MeHg content is a small fraction of the total, however, the introduced error may be acceptable. As shown in the calculations at the end of the chapter, there are some waters where MeHg is a substantial fraction of the dissolved Hg and if this is not accounted for, the error can be large. Clearly, the most accurate estimate will be obtained if the following calculation is used to derive the I-Hg fraction:

$$\underline{\text{I-Hg}} = (\underline{\text{T-Hg}} - \underline{\text{P-Hg}}) - \underline{\text{D-MeHg}} \qquad 5.4.3\text{-}4$$

where D-MeHg is the dissolved MeHg concentration. If D-MeHg is not measured directly, then it may be estimated if the K_D for MeHg is known and the total MeHg has been determined:

$$\underline{\text{D-MeHg}} = 10^6 * \underline{\text{T-Hg}}/(\text{TSS} * K_D + 10^6) \qquad 5.4.3\text{-}5$$

where TSS is in mg/L and the units of K_D are L/kg.

Other assumptions in calculating INR-Hg are needed about the nature of the ligands in solution. As a first estimate, it is reasonable to assume that this unreactive fraction consists primarily of Hg bound to NOM (Hg-NOM). As discussed in Chapter 1.0 and elsewhere, some of the Hg in solution is bound in highly stable complexes that do not degrade on the timescales of interest (days) (Hsu-Kim and Sedlak, 2003; 2005) and therefore it may be useful to separate these complexes from other complexes of Hg with NOM that are likely to be part of the "potentially bioavailable Hg pool". It is not known in detail what the form of these stable Hg complexes is, or the relative fraction of the inorganic Hg that is in this form. There is conflicting evidence in the literature whether these complexes are organically-bound Hg or inorganic Hg complexes with S-containing inorganic ligands. Neither are the sizes of these complexes known and therefore their ability to cross cell membranes is not clearly defined.

Studies of WWTP effluents suggest however that there are abundant ligands of this type in solution and that most of the dissolved inorganic Hg is likely bound to these ligands (Hsu-Kim and Sedlak, 2003; 2005). Other studies of the interaction of Hg with S in the presence of NOM suggest that there is a complex interaction between Hg, S and NOM, and that some fraction of the inorganic Hg-S species, and in particular the neutral Hg-S complexes, partition or bind with organic matter and are not bioavailable in this form (based on octanol-water partition experiments and ultrafiltration studies) (Miller, 2005; Miller et al., 2007). These studies indicate that estimations of bioavailability that do not account for these interactions severely overestimate the bioavailable fraction, and that these interactions need to be considered. Miller et al. (2007) provide the basis for making these estimates. Although a detailed description of the modeling has not appeared in the literature, it is contained in Miller's Ph.D. thesis (Miller, 2005).

In defining an approach to estimating the bioavailable inorganic Hg fraction in source waters, the results of both studies were taken into account. Thus, four different fractions or types of complexes of inorganic Hg with ligands in the filtered fraction are defined and the complexation of inorganic Hg in solution is considered to consist of: 1) simple inorganic complexes with ligands, such as $HgCl_2$ and $Hg(OH)_2$; 2) inorganic complexes formed in the presence of sulfide such as HgS, $Hg(SH)_2$ and HgS_2^{2-}; 3) organic complexes with dissolved natural organic matter (NOM) in which the Hg is binding to thiol ligands in the NOM; and 4) complexes that are a combination of neutral Hg-S complexes interacting with NOM. All inorganic complexes of type 1 are considered to be bioavailable, and only the neutral Hg-S complexes of type 2 are considered bioavailable. For the purposes of the estimation of fB-Hg, it will be assumed that the Hg-NOM fraction, either formed by simple complexation or via the Hg-S-NOM interaction (i.e. complex types 3 and 4) fall into the nonreactive inorganic Hg fraction.

The rationale for assuming that type 1 and 2 complexes are bioavailable rests on results of uptake experiments with microbes. In most instances reported in the literature, inorganic Hg uptake occurs via passive diffusion across the membrane of neutral complexes. In highly sulfidic environments, or even in oxic waters (Rozan et al., 2000) and in WWTP effluents (Adams and Kramer, 1999), there is the potential for Hg to be bound to inorganic sulfide complexes that are not readily bioavailable for uptake as they are not neutrally charged (e.g. HgS_2^{2-}, $HgSH^+$). Similarly, some of the chloride complexes are charged (e.g. $HgCl^+$, $HgCl_4^{2-}$) and may not be bioavailable in this form. However, as the ligands of the dissolved inorganic Hg species can interchange extremely rapidly, the Hg in these complexes is potentially bioavailable as depletion in the concentration of the bioavailable forms (neutral chloride and sulfide complexes) due to accumulation will result in a change in the complexation in solution to regain equilibrium. The assumption that the inorganic Hg-S complexes are inert and therefore that only the neutral complexes are bioavailable is based on limited literature about the stability of such ligands, but under essentially all the conditions that will be considered in the estimations below, the total dissolved sulfide level will be <1 μM, and under these conditions HgS is the dominant form in the dissolved inorganic Hg-S pool. Thus, while the point is theoretically important, it is not relevant to the current discussion.

In most cases in the environment, the uptake of Hg by organisms is the rate limiting step and the repartitioning of Hg between inorganic forms in solution is rapid compared to uptake; this "buffers" the bioavailable fraction in solution. Under these conditions, as discussed for example in Morel and Hering (1993), the uptake rate is related to the total inorganic Hg pool, and this is assumed in the estimations made here. It is further assumed that there are no complexes of Hg with small dissolved organic ligands, such as Hg-amino acid complexes or those with small molecular weight thiols, which could be taken up by microorganisms. Such complexes have been shown to be a small fraction of the Hg-NOM pool and have also been shown to be accounted for in the R-Hg fraction (Lamborg et al., 2003). Thus, these complexes, while organic in nature, are assumed to behave similarly to Hg-inorganic ligand complexes.

As noted above, the rate of ligand exchange for large molecular weight Hg-NOM complexes is slow and occurs on a timescale equivalent to that defined above as the reasonable timescale for the bioavailablility considerations; therefore Hg-NOM complexes are considered unavailable even through there is the potential for them to dissociate and change their speciation upon mixing. Furthermore, in terms of uptake, these large Hg-NOM complexes would not be taken up to any appreciable degree by microorganisms. Note however, that this does not apply to complexes of Hg with organic ligands such as amino acids and organic acids, which are potentially small enough, neutrally-charged, and therefore bioavailable. Addditionally, it is assumed that once Hg-S-NOM or Hg-NOM complexes form, they are kinetically inert to rapid equilibration. Thus, these do not form a part of the bioavailable Hg fraction. These complexities are inherent in Hg biogeochemistry in solution, and cannot be adequately captured in a simple ranking scheme. However, these complexities do not negate the use of a ranking scheme to provide a relative estimate of fB-Hg for various water types and sources, as portrayed in Equation 5.4.3-1. All these potentially confounding factors can be accounted for with the use of more complex modeling approaches. However, these approaches will also need much more detailed and comprehensive datasets than are currently available for their proper development.

If the NOM content of the effluent is not known or has not been measured, then it may be possible to obtain an estimate based on the TSS content. It is worth noting, given the data presented in previous chapters, that the range in NOM concentrations for WWTPs with advanced treatment is 1-10 mg/L, and that this is also the case for many other potential sources. Thus, this range would be a reasonable default value to use in the absence of data or other knowledge. Urban runoff and WWTP's without advanced treatment may have higher NOM values, however, and in such situations a more accurate assessment of the NOM concentration may need to be made.

It is worthwhile to discuss ways to obtain estimates of the other parameters that are in Equation 5.4.3-1. Clearly, gathering site specific data is the best approach in all cases, but obtaining all the information necessary to estimate all the parameters in Equation 5.4.3-1 with sufficient accuracy may be impractical. Thus, it is necessary to provide some ways to estimate all the parameters in Equation 5.4.3-1. It is assumed that T-Hg has been measured as this is needed for the estimation of many of the parameters, and for the estimation of fB-Hg.

5.4.4 Estimating the Particulate Mercury Fraction

The estimation of P-Hg (ng/L) can be done in several ways. Filtration of a known volume of the sample water (0.45 μm filter) and analysis of the filter will provide an estimate of P-Hg. Alternatively, as the Hg content of the solids in a WWTP is typically measured, it can be assumed that the solids concentration (on a mass basis) of the suspended matter in the final effluent is similar to that in the final solid removal stage (Hg_{solids} (mg/kg)). Thus, P-Hg can be estimated as:

$$\underline{\text{P-Hg}}\ (\text{ng/L}) = \underline{\text{Hg}_{solids}}\ (\text{mg/kg or ng/mg}) * \text{TSS}\ (\text{mg/L}) \qquad 5.4.4\text{-}1$$

where TSS is the suspended particulate concentration in the final effluent.

This should provide a relatively robust estimate of P-Hg. Recall the discussion above about the relationship between P-Hg and inorganic particulate Hg (IP-Hg). For WWTPs, the available data suggest that the Hg in solids, on a mass basis, is around 1 to 5 mg/kg (ppm), with many plants having higher levels in the primary solids and lower values in the final solid removal stage. The relationship shown in Equation 5.4.4-1 suggests that T-Hg concentration is a strong function of the effluent TSS. For example, given a particulate value of 1 ppm, the P-Hg concentration will increase from 1 to 20 ng/L as the TSS increases from 1 to 20 mg/L. Of course, if the TSS varies over a small range, then other factors can be as important in determining the T-Hg, such as differences in the amount of NOM in solution, or small changes in the partitioning between the dissolved phase and TSS between systems. Thus, as pointed out in Chapter 6.0, there is not always a simple relationship between T-Hg and TSS across a number of WWTPs, or across a variety of different sources to a particular waterbody.

Another approach is to use the concept of partition coefficients (K_D). These have been measured by various investigators for a number of water types, and it is possible to obtain a reasonable value for K_D from the literature. Values of K_D for WWTP effluents were estimated from available information in Chapters 2.0 and 6.0, and these values provide the typical range that has been found across most water types. The K_D has been shown to vary with the organic content of the solids and with the TSS. The variation with TSS is related to the fact that "colloidal Hg" passes through the filter and is quantified in the dissolved fraction and that the amount of the colloidal fraction is related in some manner to TSS. Thus, for waters with a high TSS, the dissolved measurement is artificially elevated by the presence of the colloidal Hg, and this leads to a lower value of K_D. From the data in Chapters 2.0 and 6.0, it appears that the $\log K_D$s range from about 5 to 6.5. The available data is plotted in Figure 5-1 and shows an average value of around 5.5. Thus, if the K_D is not known, then a range in values can be used, or an intermediate value (e.g. $\log K_D = 5.5$), which is close to the median value of the available data (Figure 5-1). With the K_D, P-Hg can be estimated, if TSS (mg/L) is known, by:

$$\underline{\text{P-Hg}}\ (\text{ng/L}) = K_D * \text{TSS} * 10^{-6} * \underline{\text{T-Hg}}/(1 + 10^{-6} * \text{TSS} * K_D) \qquad 5.4.4\text{-}2$$

where the factor of 10^{-6} is needed to convert TSS from mg/L to kg/L.

If TSS is low (1 mg/L), at a K_D of 10^5, P-Hg ~ 0.1(T-Hg), or a small fraction of the total Hg. At the other extreme, for 100 mg/L TSS, and a K_D of 10^6, P-Hg ~ 0.99(T-Hg). At a TSS of >50 mg/L, most of the T-Hg would be bound as P-Hg, even with a relatively low K_D. Therefore, the dissolved fraction can be an important part of the total Hg at low TSS, as discussed in Chapter 6.0. The relationship between K_D, TSS, T-Hg and dissolved Hg is shown in Figure 5-2. In the environment there is an overall trend that high TSS waters are likely to have a higher T-Hg. For example, combining Equations 5.4.4-1 and 5.4.4-2, you can see that:

$$\underline{\text{T-Hg}} = \underline{\text{P-Hg}}\ (\text{mg/kg})[1 + K_D * \text{TSS} * 10^{-6}]/K_D \qquad 5.4.4\text{-}3$$

At a $\log K_D$ of 5 and 1 ppm P-Hg_{solids}, T-Hg would be 11 ng/L for a TSS of 1 mg/L, while at a TSS of 10 mg/L, T-Hg would be 20 ng/L. At 50 mg/L TSS, T-Hg would be 60 ng/L. Dissolved Hg would be the same concentration under all these conditions (10 ng/L, assuming no change in K_D with TSS), so the impact of changing TSS is primarily on the T-Hg concentration and it will not have an impact on the B-Hg, as it is defined. Note that, given the discussion above about the relationship between K_D and TSS, and that shown in Chapter 6.0, there would in reality likely be some increase in dissolved Hg with increasing TSS, but there is likely

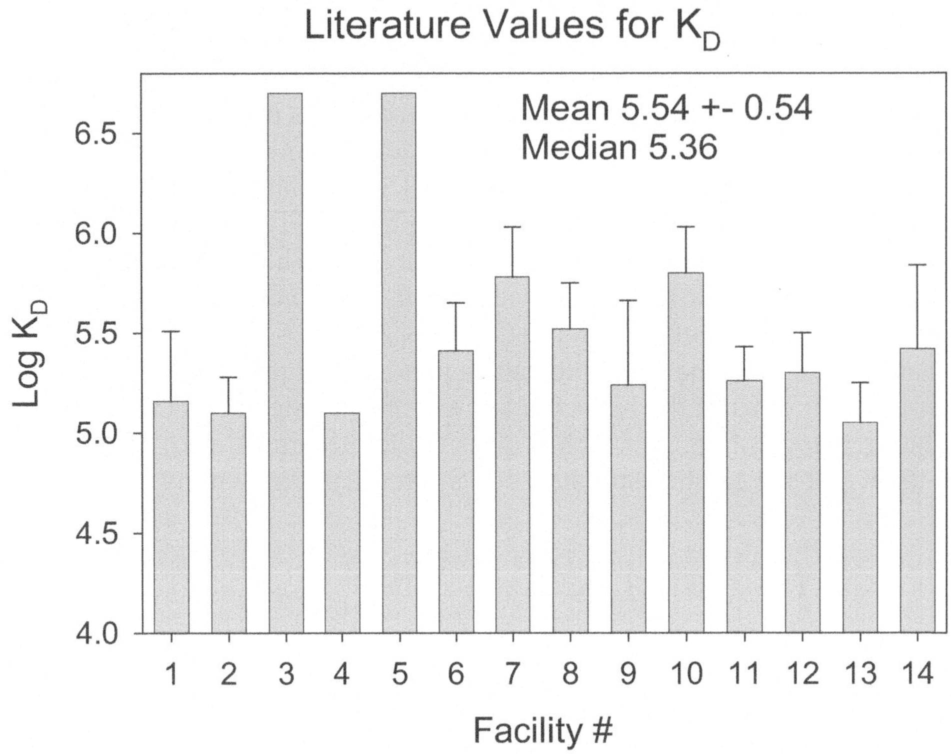

Figure 5-1. The Dissolved-Particulate Partition Coefficients Measured for a Number of Different Sewage Treatment Plant Effluents, Compiled from Data in Chapters 2.0 and 6.0.

to be little change in B-Hg, as this increase is likely due to increased colloidal Hg which is considered part of the unavailable pool.

As noted above, if K_D is not known it can be estimated but there will be error introduced with such an estimation. To examine the sensitivity of using an estimated K_D value, the estimation of fB-Hg over a range of K_D at various TSS concentrations was determined using Equations 5.4.3-1 and 5.4.4-2 under one scenario (10 ng/L T-Hg, 5% MeHg and 80% of the dissolved I-Hg as unreactive) (Figure 5-3). The result shows that the value of fB-Hg will change by a factor of about two as the K_D changes over an order of magnitude under this scenario. Thus, the fB-Hg estimate is sensitive to the value of K_D, but not in a 1:1 fashion, and there are certainly conditions where this approach may provide a reasonable first estimate of the fB-Hg.

There are more data on K_Ds and the distribution between the dissolved and solid phase for the other source profiles being considered. In these instances, it may be possible to make a more accurate assessment of the K_D value from a review of the literature. It also appears that the range of values may be lower for these other sources.

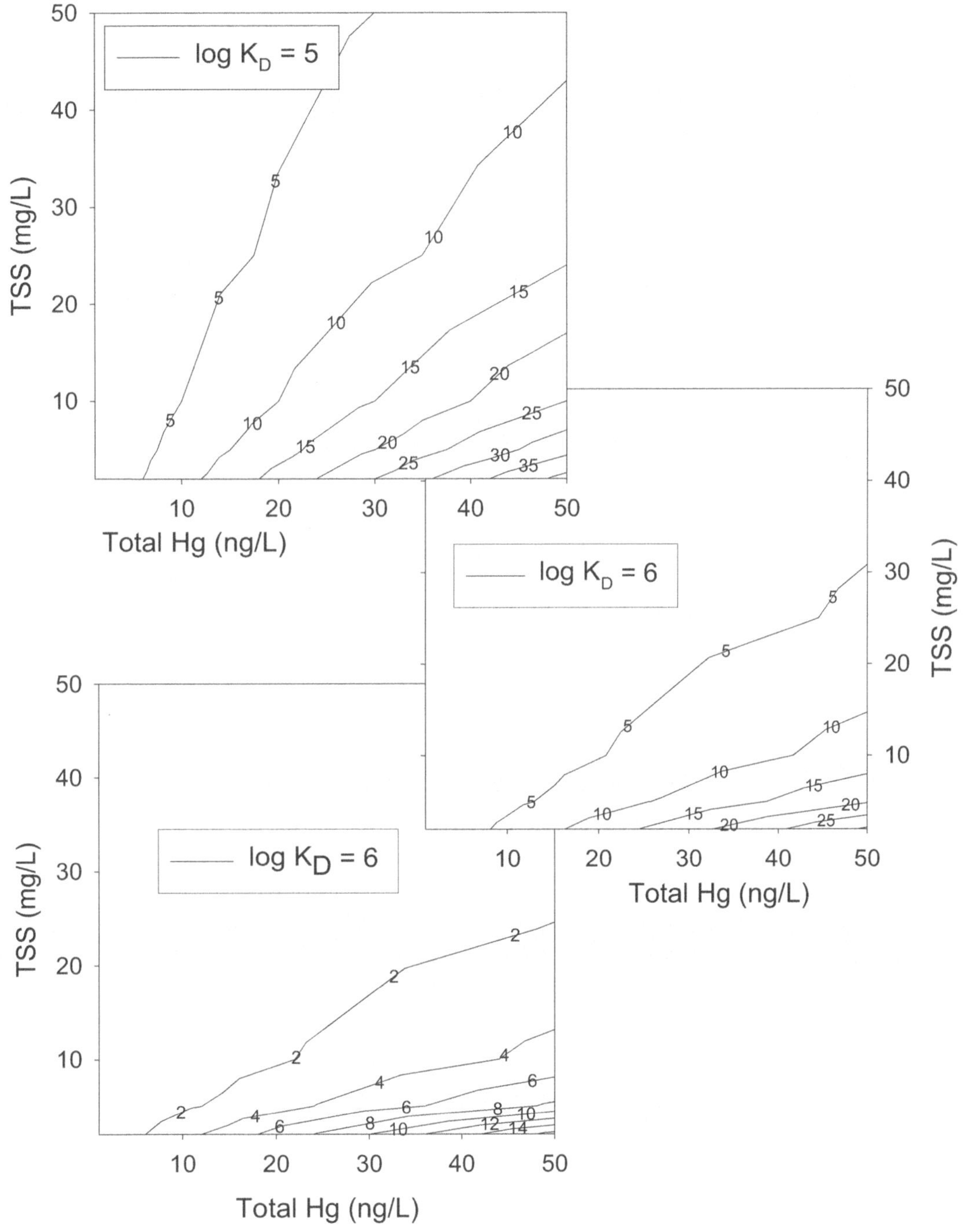

Figure 5-2. Calculated Values of Dissolved Inorganic Mercury under Different Conditions of Total Mercury, Total Suspended Solids, and Distribution Coefficients.

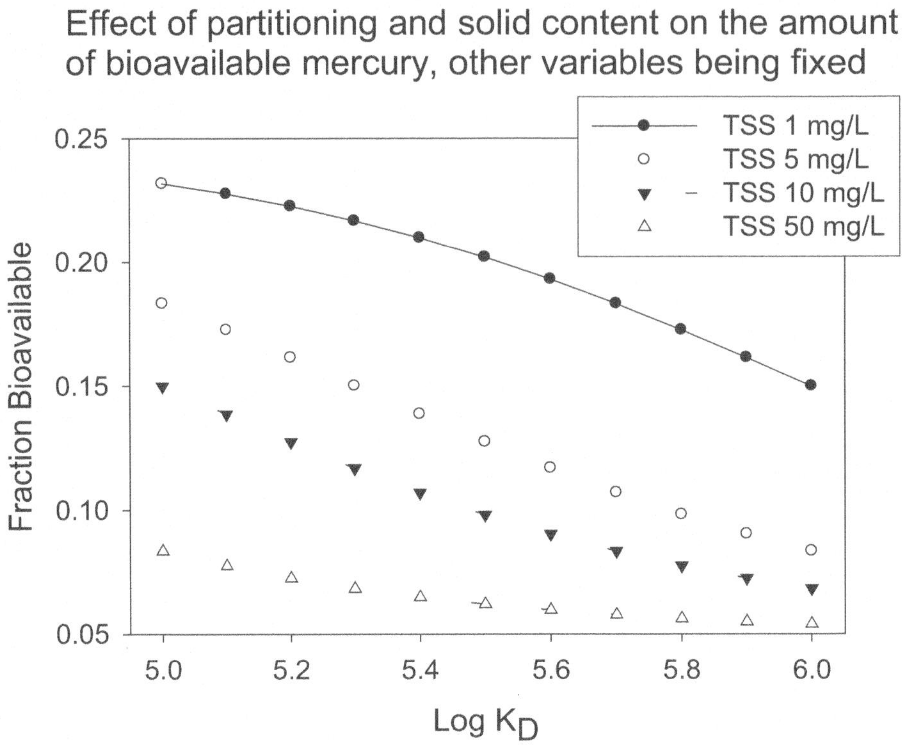

Figure 5-3. Sensitivity of the Estimation of the Fraction Bioavailable to the Magnitude of the Dissolved-Particulate Partition Coefficient and the Suspended Solids Concentration. The total mercury is set at 10 ng/L, with 5% methylmercury and 80% of the dissolved inorganic mercury bound to NOM.

5.4.5 Estimating Mercury Bound to Natural Organic Matter

For most of the sources being evaluated, the Cl concentration is insufficient to have a dominating impact on Hg speciation in the presence of NOM. However, to fully evaluate the speciation of Hg for a particular waterbody would require information on the NOM concentration, pH and the major ion (especially Cl) concentrations. For example, at pH 7, in the absence of organic matter, at a Cl concentration of 2×10^{-3} M (70 mg/L), the following complexes, in order of importance, are predicted by thermodynamic (MINEQL) calculations: $HgCl_2$ (46%), HgOHCl (43%), $Hg(OH)_2$ (8%), with other species as minor components. This relative speciation persists in the presence of organic matter, but the total fraction as inorganic species will decrease. It is necessary to make some assumptions and choose literature values for complexation of Hg to organic matter in order to predict the range of interaction between NOM and Hg. First, it is assumed that Hg is binding to NOM as a 1:2 interaction, with an organic thiol moiety on the NOM. Thus, if the NOM is represented as a thiol, RSH (where R represents the remainder of the molecule, which can dissociate into $RS^- + H^+$; $pK_a = 10$), then

$$Hg^{2+} + 2RS^- = Hg(SR)_2 \qquad 5.4.5\text{-}1$$

is the relevant reaction. The equilibrium constant from Drott et al. (2007) is used in the calculation as well as their estimation of the number of binding sites on the NOM. The number of binding sites is estimated as 10^{-8} sites per mg DOC. This value is within the range of values found by Lamborg et al. (2004) (0.9 to 4 x 10^{-8} sites per mg DOC). Thus, given the data in these papers and knowledge of the chemistry of Hg and NOM, it is concluded that 1 mg/L NOM is equivalent to ~ 2 x 10^{-8} M Hg binding sites, assuming that the ratio of NOM to DOC concentration is 0.5 (Benoit et al., 2001).

Using these constants, and the Cl concentration mentioned above, it is estimated that all the Hg will be bound to NOM even at concentrations of NOM of <0.2 mg/L at pH 7. Essentially, if there is any NOM in solution, the inorganic Hg is complexed using the estimations above. However, as noted above, it has been shown that there is dissolved sulfide in surface waters even in the presence of oxygen because the sulfide forms are kinetically inert to oxidation. Therefore it should be considered that even when sulfide is not measurable, there is likely a trace of it in solution (> 1 nM) and thus it is possible for it to have an influence on the Hg speciation, given that Hg is typically present at pM concentrations and binds very strongly with sulfide. Thus, as it has been shown that at low sulfide that the inorganic speciation is dominated by HgS (e.g. Benoit et al., 1999), then there is the possibility for the presence of bioavailable Hg even at low, unmeasurable sulfide levels.

However, the interaction found by Miller et al. (2007) also needs to be considered. The most pertinent reactions to are:

$$Hg^{2+} + HS^- = HgS + H^+ \qquad \log K_{HgS} = 26 \qquad 5.4.5\text{-}2$$

$$Hg^{2+} + 2RSH = Hg(RS)_2 + 2H^+ \qquad \log K_{NOM} = 22.1 \qquad 5.4.5\text{-}3$$

$$Hg^{2+} + HS^- + RSH + H_2O = \underline{S\text{-}Hg\text{-}RSH} + H^+ \qquad \log K_{com} = 34.7 \qquad 5.4.5\text{-}4$$

It can be seen immediately by inspection that given similar or higher concentrations of HS^- than binding sites on NOM (RSH) in typically considered environments, direct binding of Hg to NOM is not important, and a number of studies have demonstrated this (e.g. Benoit et al., 2001). However, it is apparent that the other two interactions *are* potentially important. If the equilibrium equations for reactions 1 and 3 are combined, the following results:

$$[\underline{S\text{-}Hg\text{-}RSH}]/[HgS] = (K_{com}/ K_{HgS}) * [RSH] \qquad 5.4.5\text{-}5$$

In words, the relative amount of the Hg in the bioavailable form (HgS) compared to that in the form of the complex interaction (S-Hg-RSH) depends on the NOM concentration only and not the sulfide concentration. This is only true under conditions where the sulfide speciation of Hg is dominated by HgS, but this is the case for low sulfide conditions (<1 μM) and typical pH conditions.

The computer program MINEQL was used to generate the relationship between % bioavailable Hg (essentially HgS) and NOM concentration, and the result is shown in Figure 5-4, and tabulated in Table 5-2. Note that while the discussion has been generalized in terms of NOM, most available techniques for the quantification of organic matter measure dissolved organic carbon concentrations (DOC) and therefore most information in the literature is in terms of DOC. Considering NOM to be essentially $(CH_2O)_x$ (e.g. glucose is $C_6H_{12}O_6$), then the ratio of DOC/NOM is approximately 0.4 on a molar basis (12/30). The fraction bioavailable is essentially constant across a pH gradient from 5-7 at higher DOC concentrations and from a sulfide level of 1 to 100 nM, which were the conditions tested. The value of 100 nM is consistent with the data of Adams and Kramer (1999) for the <10 kDa fraction of a WWTP effluent, but is less than the value of about 1 μM estimated by Rozan et al. (2000) for 0.4 μm filtered river waters. However, as noted above, the estimation should be valid at concentrations of sulfide below 1 μM. Although it is difficult to fit a regression curve through the data that captures all the nuances of the relationship, one possible relationship is:

$$\% \underline{\text{IB-Hg}} = 2.0 + 73.5 * \exp[-1.48 * \text{DOC}] \quad 5.4.5\text{-}6$$

where the DOC concentration is in mg/L.

This allows the direct estimation of the bioavailable fraction and the non-available fraction if the dissolved inorganic Hg concentration (I-Hg) is known or can be estimated by:

$$\underline{\text{INR-Hg}} = \underline{\text{I-Hg}} - \underline{\text{IB-Hg}} \quad 5.4.5\text{-}7$$

Table 5-2. Estimated Fraction of Inorganic Mercury that is Bioavailable for Different pH and Organic Matter Concentrations.*

DOC (mg/L)	%Bioavailable pH = 7	%Bioavailable pH = 5.5	%Bioavailable pH = 5
0.1	66.2	64.5	60.3
1.2	14.3	14.2	13.9
2.3	8.0	8	7.9
3.4	5.5	5.5	5.5
4.5	4.2	4.2	4.2
5.6	3.4	3.4	3.4
6.7	2.9	2.9	2.9
7.8	2.5	2.5	2.5
8.9	2.2	2.2	2.2
10.0	2.0	2.0	2.0
20.0	1.0		

*Values estimated using literature values for the binding constants and the computer program MINEQL.

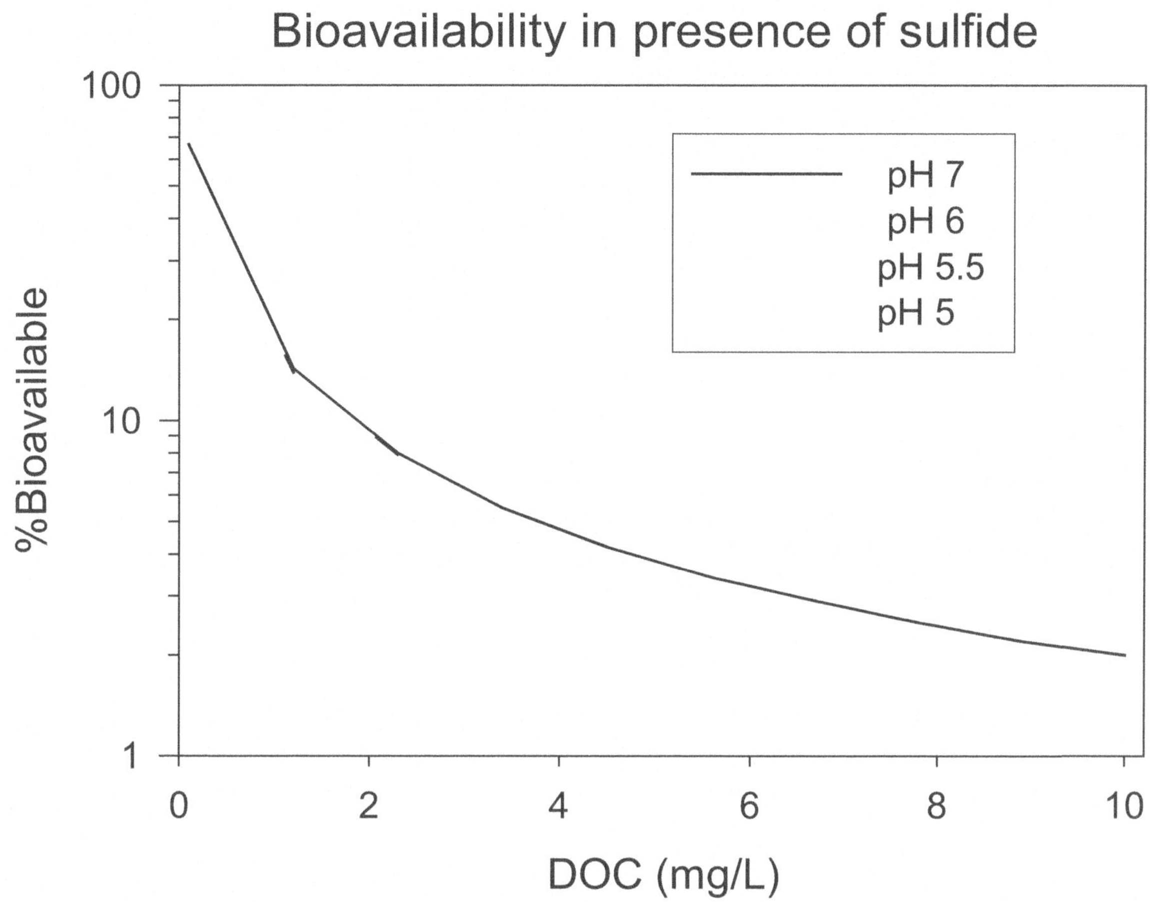

Figure 5-4. Calculated Bioavailable Fraction for the Dissolved Pool of Inorganic Mercury in Waters with Increasing Organic Matter Content (DOC) Based on Literature Values for the Interaction of Inorganic Mercury with Sulfide, Mercury with Natural Organic Matter, and the Interaction between Mercury, Sulfide and Inorganic Matter. Only Hg complexes not associated with organic matter are considered bioavailable.

Under different assumptions about the number of binding sites in NOM and the magnitude of the equilibrium constants, it is evident from the calculations in Equation 5.4.5-1 that it is possible to estimate the fraction of the Hg in solution bound to NOM as S-Hg-NOM. From the equations given above it could be possible to generate representative values across a range of pH, NOM and Cl concentrations, and other conditions. Note that there is some variability in the reported value of the binding constant for Hg with NOM and thus it is important to ensure that the correct value is used in the calculations for this association.

5.4.6 Strongly Associated Mercury Complexes

As noted above, there is the possibility that some of the dissolved Hg is complexed to ligands that bind Hg very strongly and appear to be chemically inert to ligand exchange on the timescale of days to weeks (Hsu-Kim and Sedlak, 2003; 2005). Whether these complexes would be considered inert and not bioavailable based on the definitions above is difficult to ascertain.

Due to the methods of their measurement and the fact that little is known about their molecular size, or whether they are inorganic or inorganic complexes of Hg, they may contain Hg-complexes from all the fractions identified above. Thus, until more information is available about the nature of these complexes, one must be satisfied that the definitions and estimation method above gives a reasonable representation of the bioavailable fraction.

5.4.7 Estimation Methods and Their Complexity

It is clear from the above discussion that the approaches to the estimation of relative bioavailability can range markedly in complexity, requirements for detailed chemical data, cost, and other factors. The simplest and most cost-effective approach is to use literature values or a "typical range" in values. Unfortunately, this approach is also prone to the most error. However, the estimation could be substantially improved by the measurement of a few key parameters such as total Hg and MeHg, and dissolved Hg, as well as TSS and NOM. This would allow a relatively robust estimation of fB-Hg, and an estimation of B-Hg as well. Alternatively, total Hg, MeHg and reactive Hg (R-Hg) could be measured and these parameters used to estimate fB-Hg and B-Hg, as discussed above, and this should provide a relatively robust estimate, even though the R-Hg measurement is operationally-defined. Finally, the most complex approach would be the collection of detailed information of dissolved and particulate Hg and MeHg concentrations in effluent waters, as well as needed ancillary data (NOM, POC, pH, Cl, sulfide etc), and the development of a detailed chemical speciation and fate and transport model for Hg and MeHg in the effluent.

The following steps and approaches can be used in a specific instance to determine the bioavailable Hg fraction. It is assumed that T-Hg concentration and the total suspended solid content are known as these are the minimum requirement for the determination of fB-Hg based on the discussions above.

1. If the T-Hg and TSS concentrations are the only parameters that are known, then it is necessary, and possible, to estimate the other required variables needed for the calculation (Equation 5.4.3-1).

    a. %MeHg is a required parameter and in most instances, for a particular source, an estimate could be obtained from the literature. Again, a range of values can be used, as shown in Table 5-1, for example. Or the calculation can be done without using the MeHg concentration, as described above.

    b. Additionally, it is necessary to know how much of the dissolved inorganic Hg is bioavailable. The most reasonable approach is to estimate this based on NOM (or DOC) content, as detailed above. However, this requires knowledge of the NOM content of the effluent and possibly the pH. Again, it would be possible to estimate these from the literature, or these could be obtained via measurement. In the absence of any data or literature values, an order of magnitude estimate could be obtained by assuming that the NOM content is within the range of 1-10 mg/L, as discussed above. Then, the dissolved fraction bioavailable can be estimated using Equation 5.4.5-1, giving a wide range in the bioavailable fraction (<22%). The partitioning of inorganic Hg to the solid phase needs to be estimated, as detailed above, using a literature value for the K_D, or a range of values.

    c. A first estimate can then be made using Equations 5.4.3-1 and 5.4.4-2.

d. A sensitivity analysis should be performed to assess the influence of the various estimated variables on the fB-Hg value. Such a sensitivity analysis would allow a determination of the most important parameters that need to be measured to improve the estimate of fB-Hg.

2. The sensitivity analysis would provide an assessment of the likely parameters that should be measured to improve the estimate. Improvement in the estimate of bioavailability would therefore be improved by obtaining the required information to provide a robust estimate of the following parameters:

a. %MeHg

b. dissolved and particulate Hg concentrations and the calculation of K_D

c. R-Hg

d. NOM to allow an assessment of the unreactive inorganic Hg fraction

This general approach can be further refined to take into account other secondary factors that may influence the fB-Hg estimate to a lesser degree. As an example, a dataset for a series of waters was used to estimate fB-Hg using the approach outlined above (Tables 5-3 and 5-4). In this case, the following parameters were available: total and dissolved Hg, total and dissolved MeHg, TSS, DOC and pH (Table 5-3). Some of the concentrations for D-MeHg (italics in Table 5-4) were missing and these were estimated as detailed below.

The particulate total Hg (P-Hg) and MeHg (P-MeHg) were calculated by difference and this then allowed calculation of the K_D values for T-Hg and MeHg (Table 5-4). For waters where data for MeHg was lacking, the overall average K_D for the other waters was used to back-calculate the D-MeHg and P-MeHg (all these values are in italics in Table 5-4). The % bioavailable inorganic Hg (%IB-Hg) was then determined using Equation 5.4.5-1 above, and the concentration of dissolved bioavailable Hg was estimated using this value and the concentration of D-Hg (termed IB-Hg in Table 5-4). For these waters there were instances where D-MeHg was a large fraction of the D-Hg (>20%) (Table 5-3), and therefore the calculation of inorganic bioavailable Hg was also done using the dissolved inorganic Hg concentrations (<u>I-Hg</u> = <u>D-Hg</u> - <u>D-MeHg</u>) rather than D-Hg. This value is noted as IB-Hg in Table 5-4. There are clearly differences between these two estimates and the importance of tracking MeHg concentrations into account for waters with high %MeHg is apparent.

The value of fB-Hg was determined in several ways. For Method 1, the calculations were based on Equation 5.4.3-1 and thus:

<u>fB-Hg</u> (Method 1) = <u>B-Hg/T-Hg</u>

The data show that the fB-Hg varies from 0.01 to 0.63 and its value is mostly determined by the amount of MeHg present. Thus, systems that have a high fraction of the total Hg as MeHg

Table 5-3. Data Collected from Various Waters Used in the Estimations of the Bioavailable Mercury Fraction.

WATER	TSS	T-Hg	P-MeHg	D-MeHg	D-Hg	%MeHg	pH	DOC
	mg/L	ng/L	ng/L	ng/L	ng/L			mg/L
A	4.2	2.6	0.22	0.12	0.66	8.46	7.4	7.12
B	2.3	0.33	0.034	*0.029*	0.21	10.30	6.84	2.66
C	7.75	2.74	1.59	1.1	1.89	58.03	6.49	4.14
D	2.32	0.51	0.14	0.16	0.55	27.45	7.05	10.57
E	27.9	0.6	0.37	0.15	0.45	61.67	7.58	6.79
F	6.98	0.84	0.038	*0.025*	0.54	4.52	6.98	4.31
G	7.2	0.7	0.01	0.04	0.43	1.43	7.2	2.89
H	7.3	1.15	0.21	0.09	0.47	18.26	7.3	10.44
I	7.51	0.63	0.02	*0.013*	0.45	3.17	7.51	3.77
J	7.22	0.55	0.07	*0.046*	0.34	12.73	7.22	2.03
K	4.84	6.17	0.34	0.21	0.84	5.51	4.84	9.06
L	6.71	6.71	0.04	0.09	0.38	0.60	6.71	3.38
M	7.47	1.15	0.1	0.04	0.56	8.70	7.47	8.47

Table 5-4. Calculated Concentrations, Partition Coefficients, and Bioavailable Fractions for the Waters Listed in Table 5-3.

Water	P-Hg	P-MeHg	KD Hg	KD MeHg	%IB-Hg	IB-Hg	IB-Hg	B-Hg	fB-Hg	fB-Hg	fB-Hg
	ng/L	ng/L	L/kg	L/kg		ng/L	ng/L	ng/L	Method 1	Method 2	Method 3
A	1.94	0.1	5.845	5.298	3.062	0.02	0.02	0.24	0.091	0.006	0.045
B	0.12	*0.005*	5.395	*4.849*	4.494	0.01	0.01	0.04	0.132	0.025	0.039
C	0.85	0.49	4.764	4.760	3.220	0.06	0.03	1.65	0.603	0.009	0.188
D	0.05	0.01	4.593	4.430	3.060	0.02	0.01	0.16	0.308	*-0.153*	*-0.133*
E	0.15	0.22	4.077	4.721	3.063	0.01	0.01	0.38	0.640	0.015	0.382
F	0.3	*0.013*	4.901	*4.849*	3.185	0.02	0.02	0.06	0.066	0.020	0.034
G	0.27	0.01	4.941	4.541	4.080	0.02	0.02	0.03	0.039	0.023	0.037
H	0.68	0.12	5.297	5.262	3.060	0.01	0.01	0.22	0.195	0.010	0.114
I	0.18	*0.007*	4.726	*4.849*	3.337	0.02	0.01	0.04	0.056	0.023	0.034
J	0.21	*0.024*	4.932	*4.849*	6.703	0.02	0.02	0.09	0.169	0.036	0.079
K	5.33	0.13	6.118	5.107	3.060	0.03	0.02	0.37	0.059	0.003	0.024
L	6.33	0.01	6.395	4.219	3.554	0.01	0.01	0.05	0.008	0.002	0.003
M	0.59	0.06	5.149	5.303	3.060	0.02	0.02	0.12	0.102	0.014	0.066

have high fB-Hg. In all cases, the value of IB-Hg is small, and is a small percentage of the dissolved inorganic Hg.

An alternative calculation of fB-Hg was done based on Equation 5.4.3-1:

$$\underline{\text{fB-Hg}}\ (\text{Method 2}) = (\underline{\text{T-Hg}} - \underline{\text{P-Hg}} - (\underline{\text{D-Hg}} - \underline{\text{IB-Hg}}))/\underline{\text{T-Hg}}$$

The values obtained using this method are much lower than those of Method 1, mainly due to the fact that these waters have relatively high MeHg content and the accounting method takes the MeHg concentrations into account more than once, specifically due to the fact that D-MeHg is a substantial fraction of D-Hg. With the available data it is possible to assess the impact of this, and Method 3 takes the amount of MeHg into account by:

$$\underline{\text{fB-Hg}}\ (\text{Method 3}) = (\underline{\text{T-Hg}} - (\underline{\text{P-Hg}} - \underline{\text{P-MeHg}}) - (\underline{\text{D-Hg}} - \underline{\text{D-MeHg}})(1 - \%\underline{\text{IB-Hg}}/100))/\underline{\text{T-Hg}}$$

Values estimated using Methods 1 and 3 are relatively similar and are overall consistent as fB-Hg is greater than MeHg in almost all cases – there are differences due, for example, to the D-MeHg concentration being similar or greater than MeHg, and a detection limit value of P-MeHg value being in Table 5-3. Overall, however, it appears the more information that is available, and specifically data on MeHg content, the more robust are the methods for estimating fB-Hg. It is also clear that this value can vary widely for different waters due to differences in %MeHg, in TSS and DOC, and partitioning.

It is worth mentioning here that other approaches to defining bioavailability could be used, especially if there is a concern with the estimation of the bioavailable Hg in the receiving waters after mixing. Deployment of organisms known to bioaccumulate both inorganic Hg and MeHg could be used as an indicator of the presence of, or change in the fraction of bioavailable Hg. Examples of such organisms are the Asian clam (*Corbicula*), which has been shown to accumulate Hg and MeHg rapidly from the water column. These organisms are useful even though, due to their rapid depuration of Hg, they reach steady state with their environment relatively quickly. Other potential indicator species are phytoplankton or zooplankton, both of which have been shown in laboratory and field studies to respond in a controlled way to the amount and speciation of Hg and MeHg in solution. However, to convert these relative measures of availability into actual water concentration data requires knowledge of the uptake rates and processes, and this is not often well known. Thus, it is likely that measurement of effluent water concentrations coupled with the estimation approaches discussed above for assessing bioavailability, is the likely best approach to assessing fB-Hg, and providing information that could be used for TMDL allocations, policymaking and other purposes.

5.5 Example 1: Hg Bioavailability in Four Lakes

As an illustration of the methodology presented above, it was applied to four lakes. Although these are not mercury effluents or sources *per se*, they are relevant for illustrative purposes. The water chemistry data from these four lakes was taken from EPRI (2003). Table 5-5 shows the relevant data.

Table 5-5. Data Relevant to Bioavailability Estimation for Waters of Four Lakes.

Water Quality Constituent	Pallette Lake, WI	Little Rock Reference Lake, WI	Barco Lake, FL	Lake 240, ON Canada
Unfiltered Inorganic Hg(II) (ng/L)	0.47	0.85	1.03	1.4
Unfiltered MeHg (ng/L)	0.05	0.07	0.018	0.079
DOC (mg/L)	5	3	0.85	7
pH	7.2	6.0	4.5	6.9
Chloride (mg/L)	0.3	0.3	5	0.4
Sulfate (ueq/L)	52	52	150	26
Settling Solids (TSS) (mg/L)	~1.0	~0.7	~0.8	~0.9
Particulate Inorganic Hg(II) in Littoral Sediments (ng/g)	0.3 to 2.7	76 to 191	30 to 331	0.3 to 2.7 [18]

Note some interesting comparisons among these four lakes. In particular, Lake Barco appears to be quite different limnologically from the other three lakes, which are located in the upper Midwest and lower Ontario, Canada. Lake Barco has the lowest average observed MeHg concentration, even though it has the lowest observed DOC concentration and the highest chloride and sulfate concentrations. Perhaps surprisingly, this lake has the highest fish tissue mercury concentrations. It also has the lowest pH and the highest sediment concentrations of inorganic mercury.

Calculations of bioavailable mercury in these four lakes are shown in Table 5-6. Estimated total unfiltered Hg was derived by summing the observed unfiltered inorganic Hg(II) and MeHg. Percent MeHg was then derived by dividing MeHg by the estimated total and expressing as a percent. Estimated particulate Hg(II) is calculated using Equation 5.4.4-2, the observed TSS and unfiltered Hg(II) concentrations, and an estimated $\log K_D$ of 5.5 from the literature. Dissolved inorganic Hg(II) is estimated as the difference between unfiltered and particulate Hg(II). Percent bioavailable inorganic Hg is estimated using Equation 5.4.5-1. The estimate of the fraction bioavailable Hg is derived from the sum of the %MeHg and the %IB-Hg.

Note that the inclusion of inorganic bioavailable mercury in Lake Barco raises the bioavailable fraction considerably to a level of about 20%, the remaining three lakes being in the range of 7 to 10%. This is primarily due to the low DOC value which allows more inorganic mercury to remain in bioavailable forms in solution. Even though the food chains are quite different Lake Barco and the other three lakes, the inclusion of bioavailable inorganic mercury

[18] Inorganic littoral sediment concentrations were not tabulated in the reference report. Estimating from a graphical presentation, they appeared to be in the same range as the Pallette Lake concentrations.

may contribute to an understanding of why fish tissue mercury concentrations are elevated in this lake compared to the others.

Table 5-6. Estimation of Bioavailable Mercury in Four Lakes.

Observed or Estimated Parameter	Pallette Lake, WI	Little Rock Reference Lake, WI	Lake Barco, FL	Lake 240, Ontario, Canada
Observed Unfiltered Inorganic Hg(II)	0.47	0.85	1.03	1.4
Observed Unfiltered MeHg (ng/L)	0.05	0.07	0.018	0.079
Estimated Total Unfiltered Hg (ng/L)	0.52	0.92	1.048	1,479
Observed TSS (mg/L)	1.0	0.7	0.8	0.9
Estimated particulate inorganic Hg(II) (ng/L)	0.113	0.154	0.208	0.31
Estimated dissolved inorganic Hg(II) (ng/L)	0.357	0.696	0.82	1.09
Percent bioavailable inorganic Hg(II)	2	2.86	22.9	2
Total Bioavailable fraction, fB-Hg	0.109	0.098	0.196	0.068

5.6 Example 2: Bioavailability in Mercury Sources in the San Joaquin/Sacramento Delta

The Sacramento-San Joaquin Bay-Delta Estuary was used as an example to demonstrate the estimation of the fraction of bioavailable mercury (fB-Hg) as described earlier in this chapter. Mercury from tributaries, mining runoff, wastewater treatment plant effluents, urban runoff, direct atmospheric deposition, and contaminated sediment sources were evaluated. Bioavailable mercury was estimated from each source using Equation 5.4.3-1. It would also have been possible to estimate bioavailability using Equation 5.4-1, had reactive mercury data been available. Because it was not, the exercise would have defaulted to the use of literature values. Since this has already been demonstrated, the researchers saw little value in performing the calculations again. In addition, an assessment of the variability and range in estimates for fB-Hg that result from uncertainty in the parameter values in each equation was performed. In general, the data were taken or derived from the Sacramento-San Joaquin Methylmercury TMDL document (Wood et al., 2006). However, additional sources as described below were also used for each of the following source types. All data sources, numbers of observations, and data ranges are included.

5.6.1 Tributaries

The Sacramento River was selected as a tributary, because it is the primary tributary source of Hg to the system (67% of total Hg load). The majority of the Sacramento River data were from Greene Landing, March 2000-October 2001 (Foe, 2003). These data are only a portion the Sacramento River data presented in the TMDL document (Wood et al., 2006). Although there were fewer samples, this dataset was selected because more analytes were measured and reported than in the TMDL document. The means and ranges for unfiltered total and methylmercury, and total suspended solids (TSS) were similar between the Foe (2003) and Wood et al. (2006) data. The dissolved organic carbon (DOC) range used (0.8-12.1 mg/L) was reported from monthly monitoring (11/5/86-2/5/98) at Mallard Island and was similar to the estimated range (1-10 mg/L) provided in Chapter 4.0. The pH of the river was estimated using information in Chapter 4.0; however, the pH range was also confirmed by the Sacramento surface water monitoring data collected by the Department of Pesticide Regulation.

5.6.2 Wastewater Treatment Plant (WWTP) Effluents

WWTP effluents represent 1.1% of total load to the Bay-Delta system. The SRCSD Sacramento River WWTP was selected as an example of a WWTP, because it contributes the largest estimated annual load to the system (152 g/yr) of any of the WWTPs. Only unfiltered total and methylmercury were reported by Wood et al. (2006); therefore, particulate mercury, pH, sulfate (based on the SJ/SC Facility), and chloride (based on the SJ/SC Facility) were estimated from Chapter 4.0. In addition, information in Chapter 4.0 was used to estimate the DOC range (1-10 mg/L).

5.6.3 Urban Runoff

Urban runoff contributes 1.1% of the total Hg load to the Bay-Delta system. Wet and dry weather urban runoff unfiltered total and methylmercury and TSS data were taken from Wood et al. (2006). All other data were estimated using information in Chapter 4.0.

5.6.4 Mining Runoff Sources

Mercury loading from the upstream gold mining and geothermal spring sites in the upper Cache Creek Watershed may be a potential significant source of mercury to the Bay-Delta system. Suchanek et al. (2003) measured mercury in the Harley Gulch and Sulphur Creek Mining Regions during the rainy season to estimate the maximum potential mercury loading. Unfiltered and filtered total and methylmercury, TSS, sulfate, and pH were taken from Suchanek et al. (2003) and all other data were estimated using information in Chapter 4.0.

5.6.5 Direct Atmospheric Deposition

Direct atmospheric deposition was estimated by Wood et al. (2006) to be 0.6% of the total Hg load to the Bay-Delta system. Wet deposition total and methylmercury data were taken from the San Francisco Bay Atmospheric Deposition Pilot Study reported by Wood et al. (2006). The average of the North Bay, Central Bay, and South Bay volume weighted concentrations were used. Deposition pH, and sulfate and chloride concentrations were taken from the National Atmospheric Deposition Program National Trends Network 2006 isopleth maps (only California sites used for the range). All other data were estimated using information in Chapter 4.0.

5.6.6 Sediment Pore Waters

Bioavailability of mercury in sediment pore waters was estimated using measurements reported in Gill et al. (2003). Filtered pore water measurements were available for T-Hg and MeHg.

5.6.6 Estimation of Bioavailability-Methods

As stated earlier, two different approaches are available to estimate the fraction of bioavailable mercury present at each source. The methods are distinguished by the formula used to calculate fB-Hg.

Estimation Method 1: The first method employs Equation 5.4-1. The equation is as follows:

$$fB\text{-}Hg = (MeHg + R\text{-}Hg)/T\text{-}Hg$$

where:

fB-Hg is the fraction of bioavailable Hg,

MeHg is the methylmercury concentration (ng/L)

R-Hg is the reactive mercury concentration (ng/L), and

T-Hg is the total mercury concentration (ng/L).

Estimation Method 2: The second method employs Equation 5.4.3-1. The equation is as follows:

$$fB\text{-}Hg = (T\text{-}Hg - IP\text{-}Hg - INR\text{-}Hg)/T\text{-}Hg$$

where:

fB-Hg is the fraction of bioavailable Hg

T-Hg is the total mercury concentration (ng/L)

IP-Hg is the particle bound inorganic mercury concentration (ng/L), and

INR-Hg is the non-reactive dissolved inorganic mercury concentration (ng/L).

Because there was no reactive mercury data available for these sources in this setting, only the second method was used to demonstrate example calculations.

5.6.7 Estimation of Bioavailability-Summary of Sources Investigated

The following table summarizes the values of fB-Hg calculated for each source. The individual calculations for each site are presented in detail following the summary. Following the summary, and informal ranking of the quality of the estimates is presented based on the amount and type of data that went into estimation of fB-Hg.

Table 5-7. Summary of Mercury Bioavailability (fB-Hg) by Source, Sacramento/San Joaquin Delta.

Source	fB-Hg
Sacramento River @ Green Landing	0.023
SRCSD Sacramento River WWTP	0.089
Urban Runoff-Dry	0.038
Mining Runoff – Sulphur Creek	0.006
Atmospheric Deposition	0.52
Contaminated Sediments	0.53

From this assessment, contaminated sediments and direct atmospheric deposition appear to have the highest fractions of bioavailable mercury, each at a little over 0.5. The Sacramento River WWTP appears to have the next highest fraction of bioavailable mercury approaching 9% of total. It is of interest to note the main reason that the bioavailable fraction for this source is elevated compared to some others is the relatively higher fraction of methylmercury in the effluent (about 8.7%). As noted earlier in this report, the %MeHg in WWTP effluents is more typically 2 to 3%. If this were the case here, the bioavailable fraction of the Sacramento River WWTP would be more on the order of 0.04 to 0.05, and more in line with the other tributary and urban runoff sources. This points out clearly the advantage of using locally collected data. As anticipated, the bioavailability in mining sources is lowest. Dry wearther urban runoff and tributary flows also rank fairly low, between 0.02 and 0.04. Note that some sources, the estimates of fB-Hg will be better than for others, since the amount of information available varies from source to source. After performing the calculations, the following informal quality hierarchy was applied to the assessment:

Good Estimates – Sacramento River @ Green Landing, and Mining Runoff (Sulfur Creek)
The number of directly measured parameters for these sources (including both total and dissolved fractions of total and methylmercury) allowed for the direct calculation of most of the parameters.

Fair Estimates – SRCSD Sacramento River WWTP, and Urban Runoff-dry weather measurements of total and methylmercury were the only parameters available. Other parameters needed to perform the calculations were estimated based on literature values.

Poor Estimates – Direct Deposition and Sediment Pore Waters
The only data available for this site was total mercury. All other parameters were estimated based on literature values. In addition, the value of fB-Hg for this source varied greatly as there is a wide range in the estimate of %reactive mercury in these source waters. For the sediment pore water estimates, total mercury and methylmercury in porewater (dissolved fractions-see discussion) were the only directly established values.

5.6.7.1 Overall Evaluation of Results

The mercury concentrations utilized in this exercise were the mean values taken from the TMDL document. The number of samples that represent each mean, in addition to the variation in each set of values varies greatly from site to site. As a result, the representativeness of each estimate depends in part on the accuracy of each mean.

Estimation of Bioavailability – Sacramento River @ Greene Landing 2000-2001

The information pertaining to this site includes direct determinations of total mercury, dissolved mercury, total methylmercury, and dissolved methylmercury, in addition to TSS, pH, and $SO4^-$ (Table 5-8). The amount of information available enables direct calculation of many of the parameters needed to estimate the bioavailability for this source.

Table 5-8. Sacramento River @ Green Landing Data.

Parameter	Value
Total Mercury (T-Hg)	5.77 ng/L[1]
Dissolved Total Mercury (D-Hg)	0.87 ng/L[2]
Methylmercury (MeHg)	0.119 ng/L[3]
Dissolved Methylmercury (D-MeHg)	0.048 ng/L[4]
DOC Range	0.8-12.1 mg/L[5]

[1]Source: Foe et al., CALFED Task 1A Appendix C - March 2000-Oct 2001 (Range: 2.13 - 22.89 ng/L; n = 18)

[2]Source: Foe et al., CALFED Task 1A Appendix C - March 2000-Oct 2001 (Range: 0.15 - 2.56 ng/L; n=18)

[3]Source: Foe et al., CALFED Task 1A Appendix C - March 2000-Oct 2001 (Range: 0.051 - 0.336 ng/L;n=18)

[4]Source: Foe et al., CALFED Task 1A Appendix C - March 2000-Oct 2001 (Range: 0.0185-0.090 ng/L;n=15)

[5]Source: Monitoring at Mallard Island (11/5/86 - 2/5/98)
(wwwdpla.water.ca.gov/supply/sampling/mwq/station_data/14/doc_mallardis.txt)

As the concentration of total methylmercury was directly determined, the concentration provided by the TMDL study was used. In order to complete the estimate of fB-Hg, an estimation of the reactive mercury (R-Hg) was performed.

To perform the estimation of R-Hg, the following is assumed:

$$\underline{\text{R-Hg}} \sim \underline{\text{ID-Hg}} - \underline{\text{INR-Hg}}$$

where:

R-Hg is the concentration of reactive mercury (ng/L),

ID-Hg is the concentration of dissolved inorganic mercury (ng/L), and

INR-Hg is the concentration of "non-reactive" dissolved inorganic mercury (ng/L).

Direct determinations of both dissolved total mercury and dissolved methylmercury were available, therefore the concentration of inorganic mercury were calculated by difference

$$\underline{\text{ID-Hg}} = \underline{\text{D-Hg}} - \underline{\text{D-MeHg}}$$

where:

ID-Hg is the concentration of dissolved inorganic mercury (ng/L),

D-Hg is the concentration of dissolved total mercury (ng/L), and

D-MeHg is the concentration of dissolved methylmercury (ng/L).

The percentage of inorganic mercury that is bioavailable was calculated using Equation 5.4.5-1:

$$\%\underline{\text{IB-Hg}} = 2.0 + 73.5\ e^{-1.48(\text{DOC})}$$

in which DOC is the dissolved organic carbon concentration.

DOC concentrations determined in monthly water samples collected at Mallard Island during the period from 11/5/86 to 2/5/98 give a range of DOC concentrations from 0.8 to 12.1 mg/L. Based on this range, the % bioavailable inorganic mercury are calculated for the mean and at each of the end points of the DOC concentration in an effort to determine how varying DOC concentration affects the estimated percentage of bioavailable inorganic mercury and therefore the value of fB-Hg.

Table 5-9. Calculation of fB-Hg for Sacramento River @ Green Landing.

	T-Hg = 5.77 ng/L, ID-Hg = 0.82 ng/L, MeHg = 0.119 ng/L		
	min DOC (0.8 mg/L)	max DOC (12.1 mg/L)	mean DOC (6.5 mg/L)
% bioavailable inorganic Hg	24.5	2.0	2.0
Bioavailable inorganic Hg (ng/L)	0.20	0.02	0.02
Non-bioavailable inorganic Hg (ng/L)	0.62	0.81	0.81
fB-Hg	**0.056**	**0.023**	**0.023**

The results detailed above show a significant difference in the percentage of bioavailable inorganic mercury at the two endpoints (24.5% at 0.8 mg/L and 2.0% at 12.1 mg/L). In addition, it was noted that there was no change in the percentage between the high value and the mean value (2.0% for both 6.5 mg/L and 12.1 mg/L). This is because of the exponential term in the DOC-related bioavailability calculation equation which results in small changes in the bioavailable fraction above a DOC concentration of about 5 mg/L.

In order to get a clearer idea of how the DOC concentration affects the % bioavailable inorganic mercury, the concentration range between the mean and the minimum DOC concentration was explored in greater detail. The results are presented in the following table.

Table 5-10. Sensitivity Analysis for the Effect of DOC Concentration on % Bioavailable Mercury.

	T-Hg = 5.77 ng/L, ID-Hg = 0.82 ng/L, MeHg = 0.119 ng/L					
	DOC 5.0 mg/L	DOC 3.0 mg/L	DOC 2.5 mg/L	DOC 2.0 mg/L	DOC 1.5 mg/L	DOC 1.0 mg/L
% bioavailable inorganic Hg	2.0	2.9	3.8	5.8	10.0	18.7
Bioavailable inorganic Hg (ng/L)	0.02	0.02	0.03	0.05	0.08	0.15
Non-bioavailable inorganic Hg (ng/L)	0.81	0.80	0.79	0.77	0.74	0.67
fB-Hg	**0.024**	**0.025**	**0.026**	**0.029**	**0.035**	**0.047**

These calculations that show no significant increase in the % bioavailable inorganic mercury occurs until the DOC drops below approximately 2.5 mg/L. At DOC concentrations less than this the percentage increases rapidly.

How then would this affect the overall estimation of bioavailable mercury in the Sacramento River? The sensitivity analysis shows that the value calculated for fB-Hg is low across most of the DOC range used to make the estimate. If it can be assumed that the samples used to generate the data represent a wide range of DOC concentrations (in other words, assuming that the majority of the samples do not have DOC concentrations less than 2 mg/L) the effect of the increase of bioavailable mercury on the overall value of fB-Hg would be minimal.

Estimation of Bioavailability – Mining Runoff Source (Sulfur Creek)

The information collected for Sulfur Creek includes direct determinations of total mercury, dissolved mercury, total methylmercury, and dissolved methylmercury, in addition to TSS, pH, and $SO4^-$. The amount of information available enables direct calculation of many of the parameters needed to estimate the bioavailability at this site. The DOC concentration range selected is based on information presented in Chapter 4.0 and considered to be a realistic range for DOC in this type of sample.

Table 5-11. Mining Runoff Source (Sulfur Creek) Data.

Parameter	Value
Total Mercury (T-Hg)	7031 ng/L[1]
Dissolved Total Mercury (D-Hg)	1875 ng/L[2]
Methylmercury (MeHg)	4.74 ng/L[3]
Dissolved Methylmercury (D-MeHg)	3.47 ng/L[4]
DOC Range	1-10 mg/L[5]

[1]Source: Suchanek et al., CALFED Task 5.1a Table 2 (2/21/01) (Range: 137-39,700 ng/L; n=14)

[2]Source: Suchanek et al., CALFED Task 5.1a Table 2 (2/21/01) (Range: 87-9,880 ng/L; n=14)

[3]Source: Suchanek et al., CALFED Task 5.1a Table 2 (2/21/01) (Range 0.49-20.40 ng/L; n=6)

[4]Source: Suchanek et al., CALFED Task 5.1a Table 2 (2/21/01) (Range: 0.08-14.40 ng/L; n=6)

[5]Source: Chapter 4.0

Estimation of Bioavailability (Sulfur Creek):

The calculations for fB-Hg for Sulfur Creek were conducted similarly to those for the Sacramento River @ Green Landing.

Table 5-12. Calculation of fB-Hg for Sulfur Creek.

	T-Hg = 7,031 ng/L, I-D-Hg = 1,872 ng/L, MeHg = 4.74 ng/L		
	min DOC (1 mg/L)	max DOC (10 mg/L)	mean DOC (5.5 mg/L)
% bioavailable inorganic Hg	18.7	2.0	2.0
Bioavailable inorganic Hg fraction (ng/L)	351	37.4	37.8
Non-bioavailable inorganic Hg fraction (ng/L)	1,521	1,834	1,834
fB-Hg	**0.051**	**0.006**	**0.006**

The two sources considered thus far each exhibited a similar increase in the percentage of bioavailable mercury with a decrease in DOC concentration. It was also assumed that the number of samples low in DOC (<2 mg/L) make up a proportional (and therefore small) number of the total samples for this source and the value for fB-Hg calculated using the mean of the DOC concentrations was deemed to be representative.

Estimation of Bioavailability – SRCSD Sacramento River WWTP

Only two directly measured parameters were found for this WWTP, total mercury and total methylmercury. The other parameters needed to calculate fB-Hg for this site were obtained from Chapter 4.0.

Table 5-13. SRCSD Sacramento River WWTP Data.

Parameter	Value
Total Mercury (T-Hg)	8.42 ng/L[1]
Methylmercury (MeHg)	0.73 ng/L[2]
TSS Range	2-37 mg/L[3]
DOC Range	1-10 mg/L[4]
K_D for Methylmercury Range	$10^{4.5}$-$10^{5.5}$ [5]
K_D for Total Mercury Range	10^{5}-10^{6} [6]

[1]Source: Appendix G TMDL - Table G.5b (n=302)

[2]Source: Table 6.5 TMDL; dates 12/00 - 12/01 (n=45)

[3]Source: Table 3-1, Chapter 3.0

[4]Source: Chapter 4.0

[5]Source: Table 3, Chapter 3.0 WERF WWTP Facilities

[6]Source: Figure 5-1

Estimation of Bioavailability (SRCSD Sacramento River WWTP):

The data available for this source includes direct determinations of total mercury and total methylmercury only. The value obtained for total methylmercury goes directly into the calculation of fB-Hg; however, since no dissolved fractions were analyzed, estimation of R-Hg begins with an estimation of the dissolved methylmercury fraction (D-MeHg) using the following equation

$$\text{D-MeHg} = \text{MeHg} * 10^6 / ((\text{TSS} * K_D) + 10^6)$$

where:

D-MeHg is the dissolved methylmercury concentration (ng/L),

MeHg is the total methylmercury concentration (ng/L),

TSS is the total suspended solids concentration (mg/L), and

K_D is the partition coefficient (L/kg).

The concentration of total methylmercury is the only parameter that was directly determined, and therefore two remaining two parameters (TSS and K_D) must be estimated. For the TSS concentration, the mean of the range of TSS concentrations typically found in WWTP effluent (range: 2-37 mg/L; mean 19.5 mg/L; source: Table 3-1, Chapter 3.0) was used. For the partition coefficient (K_D), the mean of the range gathered from the literature (range: $10^{4.5}$-$10^{5.5}$; mean: 10^5; source: Table 3-3, Chapter 3.0) was used.

Table 5-14. Estimation of D-MeHg for the SRCSD Sacramento River WWTP.

Parameter	**MeHg = 0.73 ng/L; TSS = 19.5 mg/L; $K_D = 10^5$**
D-MeHg	0.25 ng/L

In order to determine the sensitivity of the D-MeHg estimate to changes in TSS concentration, values for D-MeHg were calculated using the same values for MeHg and K_D, and by adjusting the TSS concentration ± 25%.

Table 5-15. Sensitivity Analysis for D-MeHg vs. TSS.

	MeHg = 0.73 ng/L; $K_D = 10^5$		
	TSS_{mean}- 25% (14.6 mg/L)	**TSS_{mean} (19.5 mg/L)**	**TSS_{mean}+ 25% (24.4 mg/L)**
D-MeHg	0.30 ng/L	0.25 ng/L	0.21 ng/L

A similar sensitivity analysis was performed by varying the value for K_D over the range taken from the literature.

Table 5-16. Sensitivity Analysis for D-MeHg vs. K_D.

	MeHg = 0.73 ng/L; TSS = 19.5 mg/L		
	$K_D = 10^{4.5}$	$K_D = 10^{5}$	$K_D = 10^{5.5}$
D-MeHg	0.45 ng/L	0.25 ng/L	0.10 ng/L

At these K_D and TSS levels, a 25% perturbation in the TSS concentration does not have have dramatic effect on the dissolved MeHg concentration. However, the value of D-MeHg is quite sensitive to the value of KD selected, underscoring the importance of this parameter in obtaining an accurate value of bioavailability.

The next step in estimating R-Hg was to determine P-Hg. In order to estimate P-Hg, the following equation was employed

$$\underline{\text{P-Hg}} = \underline{\text{T-Hg}} * K_D * \text{TSS} * 10^{-6} / (1 + (K_D * \text{TSS} * 10^{-6}))$$

where:

P-Hg is the particulate mercury fraction (ng/L),

T-Hg is the total mercury concentration (ng/L),

K_D is the partition coefficient, and

TSS is the total suspended solids concentration (mg/L).

Table 5-17. Estimation of P-Hg for SRCSD Sacramento River WWTP.

Parameter	**T-Hg = 8.42 ng/L; TSS = 19.5 mg/L; $K_D = 10^{5.5}$**
P-Hg	7.25 ng/L

The estimation of P-Hg is also dependent on the TSS concentration, and therefore the TSS concentration was adjusted ± 25% in order to analyze the sensitivity of P-Hg to this parameter.

Table 5-18. Sensitivity Analysis for P-Hg vs. TSS.

	T-Hg = 8.42 ng/L; $K_D = 10^{5.5}$		
	TSS_{mean}- 25% (14.6 mg/L)	TSS_{mean} (19.5 mg/L)	TSS_{mean}+ 25% (24.4 mg/L)
P-Hg	6.92 ng/L	7.25 ng/L	7.45 ng/L

A similar sensitivity analysis can performed by varying the value for K_D over the range taken from the literature.

Table 5-19. Sensitivity Analysis for P-Hg vs. K_D.

	T-Hg = 8.42 ng/L; TSS = 19.5 mg/L		
	$K_D = 10^5$	$K_D = 10^{5.5}$	$K_D = 10^6$
P-Hg	5.57 ng/L	7.25 ng/L	8.01 ng/L

As with the calculation of D-MeHg, the value of K_D selected has a bigger impact on the value of P-Hg than TSS.

The concentration of dissolved inorganic mercury ID-Hg was calculated from the estimates using the following equation

ID-Hg = T-Hg – P-Hg – D-MeHg

where:

ID-Hg is the dissolved inorganic mercury fraction (ng/L),

T-Hg is the total mercury concentration (ng/L),

P-Hg is the particulate mercury fraction (ng/L), and

D-MeHg is the dissolved methylmercury fraction (ng/L).

Table 5-20. Estimation of ID-Hg for the SRCSD Sacramento River WWTP.

Parameter	T-Hg = 8.42 ng/L; P-Hg = 7.25 ng/L; D-MeHg = 0.25 ng/L
ID-Hg	0.93 ng/L

The fraction of the dissolved inorganic mercury can now be estimated using the DOC concentrations as with other sources, and the value for fB-Hg can be calculated.

Table 5-21. Calculation of fB-Hg (Method 1) for the SRCSD Sacramento River WWTP.

	T-Hg = 8.42 ng/L; ID-Hg = 0.93 ng/L, MeHg = 0.73 ng/L		
	min DOC (1 mg/L)	max DOC (10 mg/L)	mean DOC (5.5 mg/L)
% bioavailable inorganic Hg	18.7	2.0	2.0
Bioavailable inorganic Hg fraction (ng/L)	0.17	0.02	0.02
Non-bioavailable inorganic Hg fraction (ng/L)	0.75	0.91	0.91
fB-Hg	**0.107**	**0.089**	**0.089**

Estimation of Bioavailability – Urban Runoff-Dry Weather Flows

Three directly measured parameters were found for this source, total mercury, total methylmercury, and TSS. Other parameters needed to calculate fB-Hg for this site were obtained from Chapter 4.0.

Table 5-22. Urban Runoff-Dry Data.

Parameter	Value
Total Mercury (T-Hg)	10.36 ng/L[1]
Methylmercury (MeHg)	0.363 ng/L[2]
TSS	16.3 mg/L[3]
DOC Range	1-10 mg/L[4]
K_D for Methylmercury Range	$10^{4.5}$-$10^{5.5}$ [5]
K_D for Total Mercury Range	10^{5}-10^{6} [6]

[1]Source: Table 7.7 TMDL (Range: 1.06-84.00 ng/L; n=59)

2Source: Table 6.11 TMDL (Range: 0.091-2.040 ng/L; n=16)

3Source: Table 7.7 TMDL (Range: 1.5-122 mg/L; n=49)

[4]Source: Chapter 4.0

[5]Source: Table 3-3, Chapter 3.0

[6]Source: Figure 5-1

Estimation of Bioavailability (Urban Runoff-Dry Weather Flows):

As with the Sacramento WWTP site, the dissolved methylmercury concentration (D-MeHg) was first estimated.

Table 5-23. Estimation of D-MeHg for Urban Runoff-Dry.

Parameter	**MeHg = 0.363 ng/L; TSS = 16.3 mg/L; $K_D = 10^5$**
D-MeHg	0.138 ng/L

In order to determine the sensitivity of the D-MeHg estimate to changes in TSS concentration, values for D-MeHg were calculated using the same values for MeHg and K_D, and by adjusting the TSS concentration ± 25%.

Table 5-24. Sensitivity Analysis for D-MeHg vs. TSS.

	MeHg = 0.363 ng/L; $K_D = 10^5$		
	TSS_{mean}- 25% (12.2 mg/L)	**TSS_{mean} (16.3 mg/L)**	**TSS_{mean}+ 25% (20.4 mg/L)**
D-MeHg	0.164	0.138	0.120

A similar sensitivity analysis was by varying the value for K_D over the range taken from the literature.

Table 5-25. Sensitivity Analysis for D-MeHg vs. K_D.

	MeHg = 0.363 ng/L; TSS = 16.3 mg/L		
	$K_D = 10^{4.5}$	**$K_D = 10^5$**	**$K_D = 10^{5.5}$**
D-MeHg	0.240	0.138	0.059

Once again, the value of dissolved MeHg was not much affected by a 25% perturbation in the TSS values, but was affected by a 10% perturbation in the $LogK_D$ value. The value for P-Hg was estimated (Table 5-26) using the median values for TSS and K_D.

Table 5-26. Estimation of P-Hg for Urban Runoff-Dry.

Parameter	T-Hg = 10.36 ng/L; TSS = 16.3 mg/L; $K_D = 10^{5.5}$
P-Hg	8.67 ng/L

The estimation of P-Hg is also dependent on the TSS concentration, and therefore the TSS concentration was adjusted ± 25% in order to analyze the sensitivity of P-Hg to this parameter.

Table 5-27. Sensitivity Analysis for P-Hg vs. TSS.

	T-Hg = 10.36 ng/L; $K_D = 10^{5.5}$		
	TSS_{mean}- 25% (12.2 mg/L)	TSS_{mean} (16.3 mg/L)	TSS_{mean}+ 25% (20.4 mg/L)
P-Hg	8.23 ng/L	8.67 ng/L	8.97 ng/L

A similar sensitivity analysis was performed by varying the value for K_D over the range taken from the literature with similar results as before.

Table 5-28. Sensitivity Analysis for P-Hg vs. K_D.

	T-Hg = 10.36 ng/L; TSS = 16.3 mg/L		
	$K_D = 10^5$	$K_D = 10^{5.5}$	$K_D = 10^6$
P-Hg	6.41 ng/L	8.67 ng/L	9.76 ng/L

The concentration of dissolved inorganic mercury I-D-Hg was calculated from the above estimates.

Table 5-29. Estimation of ID-Hg for the SRCSD Sacramento River WWTP.

Parameter	T-Hg = 10.36 ng/L; P-Hg = 8.67 ng/L; D-MeHg = 0.138 ng/L
ID-Hg	1.55 ng/L

The fraction of the dissolved inorganic mercury was then estimated using the DOC concentrations as for other sources, and the value for fB-Hg was calculated (Table 5-30).

Table 5-30. Calculation of fB-Hg for Urban Runoff-Dry.

	T-Hg = 10.36 ng/L; ID-Hg = 1.55 ng/L, MeHg = 0.363 ng/L		
	min DOC (1 mg/L)	**max DOC (10 mg/L)**	**mean DOC (5.5 mg/L)**
% bioavailable inorganic Hg	18.7	2.0	2.0
Bioavailable inorganic Hg fraction (ng/L)	0.29	0.03	0.03
Non-bioavailable inorganic Hg fraction (ng/L)	1.26	1.52	1.52
fB-Hg	**0.063**	**0.038**	**0.038**

Estimation of Bioavailability – Direct Atmospheric Deposition

Only one directly measured parameter was found for this source, total mercury. The other parameters needed to calculate fB-Hg for this site were estimated using information from Chapter 4.0 and information contained in this chapter. In order to calculate fB-Hg for direct atmospheric deposition, the ranges for percent methylmercury and percent reactive mercury drawn from the literature were used to estimate MeHg and R-Hg.

Table 5-31. Direct Atmospheric Deposition Data.

Parameter	Value
Total Mercury (T-Hg)	7.9 ng/L[1]
%Methylmercury (MeHg)	0.5-6%[2]
% Reactive Mercury (R-Hg)	15-95%[3]
DOC Range	1-10 mg/L[4]
K_D for Methylmercury Range	$10^{4.5}$-$10^{5.5}$ [5]
K_D for Total Mercury Range	10^{5}-10^{6} [6]

[1]Source: : Table 7.10 TMDL (Volume Weighted-North Bay: 7.4 ng/L (n=14); Central Bay: 6.6 ng/L (n=16) ; South Bay: 9.7 ng/L (n=29))

[2]Source: Based on the MeHg (Range: 0.5 - 6%) Chapter 4.0

[3]Source: Table 5-1

[4]Source: Chapter 4.0

[5]Source: Table 3, Chapter 3.0

[6]Source: Figure 5-1

Table 5-32. Estimation of Total Methylmercury for Direct Atmospheric Deposition.

	T-Hg = 7.9 ng/L		
	0.5%	**3.25%**	**6.0%**
MeHg	0.04 ng/L	0.26 ng/L	0.47 ng/L

An estimation of reactive mercury (R-Hg) was also made from the ranges provided in Chapter 4.0.

Table 5-33. Estimation of Reactive Mercury for Direct Atmospheric Deposition.

	T-Hg = 7.9 ng/L		
	15%	**55%**	**95%**
R-Hg	1.19 ng/L	4.35 ng/L	7.51 ng/L

Not surprisingly, due to the large range for percent reactive mercury, the accuracy of the final estimation of fB-Hg varies widely. In order to gauge the sensitivity of fB-Hg to the estimate of methylmercury concentration, the corresponding value for %R-Hg was estimated at the minimum value of 15%. The higher the percentage of R-Hg, the less impact on fB-Hg from the methylmercury concentration.

Table 5-34. Estimation of fB-Hg with Varying %MeHg.

	T-Hg = 7.9 ng/L, %R-Hg = 15%		
	0.5%	**3.25%**	**6.0%**
fB-Hg	0.16	0.18	0.21

Thus, the value for fB-Hg can vary quite a bit (there is approximately a 30% difference between the endpoints) if the percentage of reactive mercury in the sample is low. While it is clear that fB-Hg will vary depending on the actual value, it is also worthwhile to see how it varies with smaller changes in %R-Hg by adjusting the value ± 15% around the mean.

Table 5-35. Estimation of Reactive Mercury for Direct Atmospheric Deposition.

	T-Hg = 7.9 ng/L; %MeHg = 3.25%		
	40%	**55%**	**70%**
fB-Hg	0.38	0.52	0.66

Since the reactive mercury is defined to be bioavailable, any variation in the %R-Hg will result in large differences in fB-Hg.

Estimation of Bioavailability – Sediment Pore Waters

Two direct measures were available for this portion of the assessment, total mercury in pore water and methylmercury in pore water. Since filtration of the sample is a critical step in pore water collection methods, both the concentrations for total and methylmercury represent the dissolved fraction (no unfiltered total mercury or unfiltered methylmercury data are available).

The other parameters needed to calculate fB-Hg for contaminated sediments were estimated using information from Chapters 3 and 4 of the document detailing the methodologies for estimating bioavailability.

Table 5-36. Sediment Pore Water Data.

Parameter	Value
Total Mercury in Porewater(Hg_D)	4 ng/L[1]
Methylmercury in Porewater ($MeHg_D$)	2.1 ng/L[2]
Sulfide (SO_4^{-2}) Range	100-1,000 mg/L[3]
Chloride (Cl^-) Range	100-10,000 mg/L[4]
DOC Range	10-100 mg/L[5]

[1]Source: Gill et al., CALFED Task 4B Figure 3- calculated mean of min and max (min: 1 ng/L, max: 8 ng/L)

[2]Source: Gill et al., CALFED Task 4B Figure 4- calculated mean of min and max (min: 0.2 ng/L, max: 4 ng/L)

[3]Source: Table 4-8, Chapter 4.0

[4]Source: Table 4-8, Chapter 4.0

[5]Source: Table 4-8, Chapter 4.0

Estimation of Bioavailability – Sediment Pore Waters:

The value of methylmercury in porewater from the TMDL document was used to represent fraction of bioavailable methylmercury in Equation 5.4-1. The directly determined values for dissolved total mercury and dissolved methylmercury in porewater can be used to calculate the dissolved inorganic fraction.

Table 5-37. Calculation of ID-Hg for Sediment Pore Waters.

Parameter	**D-Hg = 4 ng/L; D-MeHg = 2.1 ng/L**
ID-Hg	1.9 ng/L

The fraction of the dissolved inorganic mercury was estimated using the DOC concentrations as with other sites, and the value for fB-Hg was calculated (Table 5-38). As demonstrated by the sensitivity analysis represented in the table below, no change in the bioavailability was indicated over the range of DOC concentrations.

Table 5-38. Calculation of fHg_B for Sediment Pore Waters.

	D-Hg = 4 ng/L; ID-Hg = 1.9 ng/L, D-MeHg = 2.1 ng/L		
	min DOC (10 mg/L)	**max DOC (100 mg/L)**	**mean DOC (55 mg/L)**
% bioavailable inorganic Hg	2.0	2.0	2.0
Bioavailable inorganic Hg fraction (ng/L)	0.04	0.04	0.04
Non-bioavailable inorganic Hg fraction (ng/L)	1.86	1.86	1.86
fB-Hg	0.535	0.535	0.535

Based on this estimate, approximately half of the mercury concentration is bioavailable. This is not surprising as it was noted that the methylmercury in porewater concentration is approximately 50% of the total mercury. While the bioavailable fraction for sediment pore waters presented is significantly higher than most of the other estimates performed, this estimate suffers from the paucity of direct information available.

5.7 Bioavailability after Mixing with Receiving Waters

As noted above, while there may be a relatively low fB-Hg in a particular effluent or input to receiving waters, especially in comparison to the fB-Hg of other sources, the overall input of B-Hg may still be high if the input (flux) of water is relatively large. If there is a need to compare two sources with similar flow rates, then clearly it is sufficient to compare the bioavailability of these two waters directly as their relative impact on mixing with the receiving waters will depend mostly on their B-Hg. However, if one input has a much higher water flow, then this needs to be considered in the determination and assignment of relative inputs of bioavailable Hg.

If the effluent or input flow is small compared to receiving water volume/flow rate, then the receiving water characteristics will have the largest impact on the relative bioavailability in the receiving waters, and the chemistry of these waters, if they are substantially different from that of the effluent, should therefore be used to modify the estimate of fB-Hg upon mixing, taking into account the caveats and discussions above. The modification of the fB-Hg estimates for the effluent could also be done using a concentration-based approach, by estimating B-Hg for both the source waters and the receiving waters and assuming that the initial mixing is conservative in terms of B-Hg. This is likely a conservative approach given the discussions in Chapter 2.0 about the impact of changes in water column characteristics, flow regimes and other factors on the partitioning and stability of Hg in solution. Also, this assumes that there is no demethylation immediately upon mixing of these waters, and that there is not rapid transport of the added Hg to the sediment where it may be incorporated into reduced oxygen environments that may stimulate methylation. Given the variability in the size and type of receiving waters to which WWTP effluents are discharged, there are obviously a large number of potential scenarios. The investigation of these will be dealt with in detail in Phase II of this project.

APPENDIX A

REMOVAL OF MERCURY BY WASTEWATER TREATMENT PLANTS: A LITERATURE AND DATA REVIEW

This appendix addresses the removal of mercury by wastewater treatment plants. It is important to note that there have been relatively few data collected to support this type of analysis. It is only recently, and then only because of regulatory drivers such as mercury TMDLs, that measurements of total mercury in wastewater effluents have been measured with low enough detection limits to establish concentrations quantitatively have been made. With the discovery of low levels of mercury in wastewater effluents, some very limited studies have been initiated to establish mercury levels in raw sewage and other influents to wastewater treatment plants. Even fewer studies have been performed look at actual wastewater treatment processes with the intent of determining the fate of mercury within a wastewater treatment plant.

Similarly, very few studies have been performed to determine MeHg levels in either wastewater influents or effluents. With the discovery of low levels of MeHg in wastewater effluents, and the knowledge that biological processes are utilized in these plants for waste treatment, a concern was raised that MeHg might be produced in wastewater treatment plant environments.

Finally, little or no work has been done with regard to the effect of treatment processes on the *bioavailability* of mercury (i.e. the effect of treatment on the bioavailable fraction in an effluent), including the presence or absence of potentially bioavailable forms or of bioavailability mitigators and enhancers. Some studies of this type are underway. However, this report represents the first effort of which the authors are aware to attempt an evaluation of these effects. Because data is limited or lacking in this regard, the team's results necessarily focus on uncertainties rather than certainties. Many of their conclusions are deduced from anticipated physical, chemical or biological reactions rather than actual data.

Appendix A is organized into six subsections. The first, Section A.1, is a summary of the key findings of this portion of the work. Section A.2 provides a brief review of the limited studies of mercury fate in wastewater treatment plants. Section A.3 reviews information obtained as a result of data requests made through WERF for this project. Section A.4 provides a discussion of existing and emerging technologies for mercury removal from wastewater. This section reviews several technology review papers, which tend to deal with more established technologies, as well as recent individual contributions from the literature, which tend to focus on emerging technologies. Section A.5 reviews information obtained on treatment technology cost. Finally, Section A.6 provides a discussion of the effects of both existing and prospective wastewater treatment technologies on the bioavailable fraction of mercury in municipal wastewater effluents.

A.1 Summary

In summary, the following conclusions with regard to wastewater treatment on mercury in municipal effluents are drawn:

- Mercury is very effectively removed by wastewater treatment plants. Depending on the treatment technologies employed, and the influent concentration, from 80% to greater than 99% of the mercury entering a wastewater plants is removed. Because of the affinity of mercury for particulates, most of the mercury becomes incorporated in biosolids. If the biosolids are incinerated as part of the treatment process, much of this mercury may be lost to the atmosphere through volatilization.
- Some promising technologies are emerging for advanced treatment of mercury following secondary treatment; however, there continues to be a dearth of information regarding the testing of these technologies at the low levels found in currently operating WWTPs and with municipal wastewaters. At this time, there are no proven, cost-effective methods to bring typical total Hg levels in municipal wastewater effluent (mostly 10 to 100 ng/L) to typical regulatory levels (1 to 5 ng/L). The most promising of these emerging technologies are either chemical or biological processes (i.e. addition of chemical reductants, photoreduction, and microbial reduction) that involve reduction of Hg^{2+} to Hg^0 and the subsequent volatilization or stripping of Hg^0 from the aqueous media. However, the loss of this Hg to the atmosphere may not be desirable from a regulatory standpoint. The costs of these technologies are not well-established; however, the cost of biological treatment appears to be relatively high. Furthermore, most of these technologies have been evaluated on industrial wastewaters having very different characteristics from municipal effluents, and at much higher T-Hg concentrations. As such, any technologies of interest should be tested on municipal effluent and at lower concentrations to properly evaluate treatment potential and cost.
- In general, wastewater treatment plants appear to generate only small quantities of MeHg as part of the treatment process and overall substantially reduce MeHg in effluent relative to influent. However, some plants appear to have effluent MeHg levels that are elevated with respect to background. Potential source areas for MeHg in these plants and identification of control opportunities should be investigated.
- Many primary and secondary wastewater treatment processes would appear to enhance the bioavailable fraction of mercury even though, overall, wastewater treatment is very efficient at removing total mercury (as well as bioavailable mercury) from the effluent stream[19] Enhanced bioavailability is deduced from the tendency of the treatment processes to promote a higher percentage of dissolved Hg in effluents relative to influents. However, this potential effect may be strongly mitigated by the presence of both organic and inorganic ligands which bind dissolved Hg strongly and may greatly reduce its bioavailability[20]. Advanced treatment

[19] The reader is reminded of definitions adopted by the authors in this context. Usage of the term "bioavailability" connotes the concentrations of bioavailable species relative to the total mercury concentration in a source or effluent. It is construed by the authors to have the same meaning as "relative bioavailability" and may be used interchangeably in the text. The term "bioavailable mercury" on the other hand connotes the actual concentrations of bioavailable mercury species in a sample. Thus, an effluent may have more "bioavailable mercury" (i.e. higher concentrations) in comparison to another source, but have lower "bioavailability" (i.e. bioavailable fraction).

processes that target weakly complexed Hg in solution are presumed to decrease Hg bioavailability.

- The effects of wastewater treatment technologies on the bioavailability of mercury are not well-established and data are lacking in this regard. More information needs to be generated in the following areas:
  - the distribution of mercury in effluents in either the dissolved or adsorbed phase (fraction of dissolved versus particulate mercury),
  - the degree of strong/weak complexation of dissolved mercury by organic and inorganic ligands,
  - the fraction of MeHg in effluents, and
  - where, and the degree to which, recalcitrant ligands form complexes with mercury and opportunities to alter such complexation.

A.2 Fate of Mercury in Wastewater Treatment Plants

Very few studies of the fate of total mercury in wastewater treatment plants have been performed. Only five studies were identified that had measured mercury in influents and effluents. Balogh and Liang (1995) sought to track the fate of mercury in the plant. In their study, they found that approximately 95% of the mercury entering the plant was accumulated in wastewater sludges and was lost to volatilization during the incineration process. Since that time, four additional studies of mercury fate in WWTPs have been undertaken. The first was for the Onandaga METRO WWTP, previously discussed in Chapter 3.0. The San Jose/Santa Clara WPCP ESD (2007), also performed a study of total and methylmercury fate and transport in their plant, as reported in Chapter 3.0. The fourth was a study recently completed for the Southerly Wastewater Treatment Center (WWTC) near Cleveland, OH (NEORSD 2006), reported below, and the fifth was a study of total and methylmercury fate and transport in at the Sacramento County Regional Sanitation District (SCRSD), also reported below.

A.2.1 Southerly WWTC Total Mercury Study

The Southerly WWTC has an average flow of 122 MGD and discharges effluent to the Cuyahoga River. The plant operates a two stage activates sludge treatment system, with the first stage being designed to remove carbonaceous BOD and the second for nitrification. Effluent from the second stage is filtered and chlorinated before being discharged. Sludges from the activated sludge units are thickened, dewatered and incinerated. Incinerator scrubber effluent and waste from the ash lagoons are recycled to the headworks of the plant.

---

[20] As was pointed out in Chapter 3.0, municipal wastewaters appear to have metal sulfide:NOM complexes that strongly bind Hg and are not removed by filtration. However, there is debate as to the composition of these complexes and there is at present no reliable information as to their effect on the bioavailability of bound mercury. To the extent that they consist of larger DOC:sulfide complexes, they may reduce bioavailability. To the extent that they tend to be simple soluble, neutral inorganic HgS, they may be readily bioavailable. However, evidence presented in earlier chapters regarding the effects of similar complexes on the toxicity of silver and copper leads us to believe that these complexes of Hg would tend not to be readily bioavailable. Obviously, there is much needed research on this topic.

The mass balance showed that the average total mercury loading attributable to the raw influent during the period of study was 61.0 grams/day. An average mercury load of 123.9 grams/day from the Easterly Pumped Sludge (EPS) entered the SWWTC treatment process at the gravity thickeners. Therefore, the sludge going to the incinerators, incorporating both the raw influent and the EPS, had an average total mercury loading of 184.9 grams/day before subtracting the effluent loading. The majority of mercury was lost through the incineration process and released to the atmosphere (152 g/day or about 82%).

Although effluent concentrations were apparently not measured during the mass balance study, effluent concentrations were measured as part of a study of the Cuyahoga River for the same plant (NEORSD 2006a). Measured over the course of five sampling events from July 20, 2004 to April 19, 2005, effluent concentrations averaged 2.3 ng/L. During three sampling events which occurred during the mass balance study (July 6, August 2, and September 13, 2005), influent concentrations ranged from 91.9 ng/L to 844 ng/L with an average of 219 ng/L. Based on these measurements, admittedly not taken concomitantly, the removal rate for mercury for the Southerly plant is estimated to be 98.9%. The Cuyahoga River study also showed that the percentage of dissolved mercury discharged ranged from 40 to 69% with an average of 56%. MeHg measurements were not made. Other ancillary measurements made of the effluent during the Cuyahoga River study are shown in Table A-1.

Table A-1. Ancillary Water Quality Parameters Measured in the Southerly WTTC Effluent from July 20, 2004 to April 19, 2005. Data Source: NEORSD (2006a).

	DOC (mg/L)	TSS (mg/L)	pH	DO (mg/L)	Conductivity (uS/cm)
Minimum	11	2	6.87	4.29	961
Maximum	13.7	6	7.66	9.23	1266
Average	12.4	3.2	7.21	7.1	1097.5

T-Hg in the effluent was negatively correlated with DOC (r = -0.96, n=4, *p=0.05*), TSS (r = -0.80, n=5, *not significant*) and somewhat negatively correlated with pH (r = -0.47, n=4, *not significant*)[21]. The ratio of dissolved to total Hg was positively correlated with pH (r = 0.62). The remaining correlations of ancillary parameters to either T-Hg, dissolved Hg, or the ratio of dissolved:total Hg were weak. The negative correlation of TSS to T-Hg is surprising as positive correlations were observed by Hsu (2006) and it is generally thought that T-Hg increases in effluents with TSS, as mercury is strongly adsorbed to particulates. However, this may just be a function of the small sample size. Based on these data, the average K_d for mercury in the discharge is 2.49E5 L/kg or $\log K_D = 5.4$.

[21] Given the small number of data points, only the T-Hg:DOC relationship is statistically significant (at the 95% confidence level) and extrapolation of trends from this small dataset is not recommended. The researchers point these relationships out, however, because they do reveal trends which are counterintuitive and show that positive relationships between T-Hg:TSS and T-Hg:DOC should not necessarily be assumed.

A.2.2 SCRSD Total and Methylmercury Fate and Transport Study

The Sacramento Regional County Sanitation District performed a study to investigate the fate and transport of MeHg throughout its treatment process and recycle streams. This study was conducted from September 2004 through May 2005 at the Sacramento Regional Wastewater Treatment Plant (SRWTP). The SRWTP uses conventional primary treatment followed by high purity oxygen secondary treatment, chlorine disinfection, and discharges treated effluent to the Sacramento River in the town of Freeport, south of the city of Sacramento.

The study utilized a two-phased approach; Phase 1A and Phase 1B. Phase 1A focused on the principal liquid train locations and return flows. Phase1B sampling plan was developed from the results of Phase 1A and focused on data needed to investigate MeHg production and removal mechanisms in more detail. The study's main objectives were to evaluate the fate and transport throughout the plant, and to identify key factors that affect MeHg. Results concluded that approximately 75% of MeHg in the influent was removed by the treatment process, with the majority or the removal occurring in the secondary treatment process. Approximately 98% of total mercury (T-Hg) was removed by the treatment process. The key difference between the two mercury forms is their particle affinity. The partition coefficient for MeHg ($\log K_D \approx 4.0$) is approximately ten-fold lower than for T-Hg ($\log K_D \approx 5.0$) as found in most environmental waters. Mass transport models show that this attribute is the dominant factor controlling the difference in removal efficiencies between MeHg and T-Hg at the SRWTP, and likely the case for other facilities. Methylation maybe occurring in the anaerobic digesters, as is evident by higher concentrations of MeHg in particles in the digesters and solids storage basins. Return flows from the on-site Biosolids Recycling Facility and Solids Storage Basins increase the influent MeHg loading by approximately 13%.

Average concentration of MeHg in the influent during the Phase 1A study was 1.56 ng/L. Concentrations and resulting removal efficiencies after primary treatment, secondary treatment, and in the final plant effluent were 1.38 ng/L (11.5%), 0.489 ng/L (68%), and 0.665 ng/L (57.4%), respectively in the Phase 1A study. Removal efficiencies for MeHg in the Phase 1B study averaged 11% after primary treatment, 75% after secondary treatment, and 70% in the final effluent. In terms of T-Hg, 43% removal occurred on average during primary treatment, 94% had occurred after secondary treatment, and 98% of the influent T-Hg had occurred before final discharge.

Ratios of MeHg to T-Hg were 0.8% in the influent, rising to 2.6% after primary treatment, 7.7% after secondary treatment, and were about 12% in the final plant effluent. The ratio of dissolved to total MeHg remained fairly constant throughout the treatment process ranging from a low of 38% after secondary treatment to a high of 46% in the final effluent. In terms of T-Hg, there was a trend of increasing percentages in the dissolved phase, increasing from a low of 1.6% in the plant influent to a high of 50% in the final plant effluent.

A.3 Results of WERF/NACWA Data Request

The National Association of Clean Water Agencies (NACWA) examined influent and effluent mercury and TSS levels in 12 WWTPs over a three year period from July of 2003 to July of 2006 (NACWA 2007). While the NACWA study was aimed at determining the efficacy of controlling sources of mercury to the plant, such as dental clinics, it also provided valuable information for this study. Influent T-Hg mercury levels were found to be highly variable, ranging over two orders of magnitude from a little greater than 10 to over 1000 ng/L. The report states that the variability in influent concentrations is most likely due to the inconsistent nature of the influent and the fact that, because the samples have varying amounts of solid particles, a truly representative influent sample is virtually impossible to collect. In light of this inherent variability, the ability to use influent mercury concentrations to determine impacts from potential mercury reduction techniques in the collection system is limited.

Effluent levels exhibited similar variability, ranging from less than 5 to over 50 ng/L. Effluent mercury concentrations exhibited variability between plants, but less variability within plants. Unlike the influent concentrations, the effluent mercury concentrations were more consistent at individual facilities. The decreased variability within individual POTWs indicates that, generally speaking, each individual POTW is consistently performing to the same level. No information on MeHg or reactive mercury levels were collected as part of the study.

While T-Hg removal for all plants was generally greater than 90%, the percentage of dissolved Hg in effluent varied widely from plant to plant. At eight WWTPs, it was reported that there appeared to be a direct relationship between TSS and T-Hg in effluent, while at the remaining five plants, this relationship was not evident in the data[22]. Sewer cleaning projects that occurred during the sampling period were found to be positively correlated with higher T-Hg levels. Factors than were correlated with lower mercury levels included higher percentages of industrial flows to the WWTP and the use of iron salts ($FeCl_3$) during the treatment process. Factors that were unrelated to mercury levels were plant flow rate, population density, whether the WWTP accepted hauled-in waste, or whether the collection system was comprised of combined or separate sewers.

A.3.1 Dissolved Hg Levels in WWTP Effluent

In terms of the current study, the most relevant information was the data gathered on dissolved levels of T-Hg in WWTP effluents. This information was gathered on a total of nine plants. Median percent dissolved T-Hg in the effluents of these plants ranged from 22% to 59%. The plant with the highest dissolved to total Hg ratio was a tertiary treatment plant (WWTP "G"). At this plant, the secondary treatment consists of high-purity oxygen activated sludge with secondary clarification. This is followed by flocculation tanks and dual-media filtration. The high percentage of dissolved T-Hg stands to reason, as the tertiary treatment processes involved tend to focus on solids removal. Of the nine plants at which dissolved T-Hg levels were measured, this plant was in the top three performing plants for TSS in effluent (average of 5.0 mg/L) and was the best performing plant for percent TSS removal (98%).

The next highest median dissolved percent T-Hg levels (~ 42%) occurred at WWTP "C". At this plant, secondary treatment is employed consisting of trickling filters, solids contact tanks, aeration and final settling. Scrubber water from the sludge incinerators is reintroduced into the influent at this plant. This plant was the second best performing plant for effluent TSS (4.3 mg/L) and in the top three best performing plants for %TSS removal (96%).

The lowest median percent dissolved T-Hg levels occurred at WWTP "A". The treatment processes at this plant consist of primary treatment followed by aeration and secondary clarification. Thus, it makes sense that it should be one of the worst performing plants among the nine in terms of solids removal. In fact, it ranked ninth in terms of effluent TSS (average of 15.7

[22] Again, these findings point out that a positive T-Hg:TSS relationship cannot always be assumed.

mg/L) and in terms of percent TSS removal (92%). The plant with the next lowest median dissolved T-Hg levels (WWTP "H") ranked surprisingly high with regard to effluent TSS - 4th with an average of 6.42 mg/L, and with regard to TSS removal efficiency – 2nd at 97.7%. Solids from the secondary clarifiers are recycled to the beginning of the treatment process at this plant; however, this plant also ranked 4th best out of the twelve in terms of overall removal efficiency. The plant with the third lowest median dissolved/total ratio (WWTP "J") was more similar to WWTP "A" in terms of solids removal performance with the 8th ranked TSS effluent levels (15.4 mg/L average) and 8th ranked TSS removal performance (95%).

Although there are departures from the rule, it appears that WWTPs that remove more solids tend to have higher dissolved/total Hg percentages in the effluent and poorer performing WWTPs for solids removal tend to have lower dissolved/total Hg percentages. This makes sense in terms of what is known about Hg partitioning. Based on data in the NACWA report, the best performing plants for solids removal have dissolved to total percentages in the 40 to 60% range, whereas the worst performing plants for solids removal tend to have dissolved/total percentages in the 20 to 30% range. The NACWA reports notes that there was no apparent relationship between higher levels of dissolved Hg in the plant effluent and the recycling of incinerator scrubber water[23].

It is of interest to note that there appears to be a relationship between T-Hg and dissolved Hg in the effluents from these WWTPs[24]. Correlation coefficients between filtered and unfiltered T-Hg ranged from 0.16 to 0.97. However, the correlation coefficient between filtered and unfiltered T-Hg when all effluents were considered was 0.87 ($R^2 = 0.761$). This indicates that when unfiltered T-Hg in the effluent is high, dissolved T-Hg also tends to be high. Figure 6-1 shows the overall relationship between filtered (dissolved) and unfiltered T-Hg for these nine WWTPs.

---

[23] These relationships suggest a "colloidal effect" in that WWTPs that perform well with regard to TSS removal likely employ other processes that also remove colloidal (or DOC-bound) Hg to the solids. Thus those plants that are efficiently removing solids are also efficiently removing colloids and the associated Hg.

[24] This stands to reason as the K_Ds for these plants are reasonably similar: D-Hg/T-Hg = $10^6/(K_D.TSS + 10^6)$, which is essentially constant except for very high TSS (in mg/L) or highly variable K_Ds.

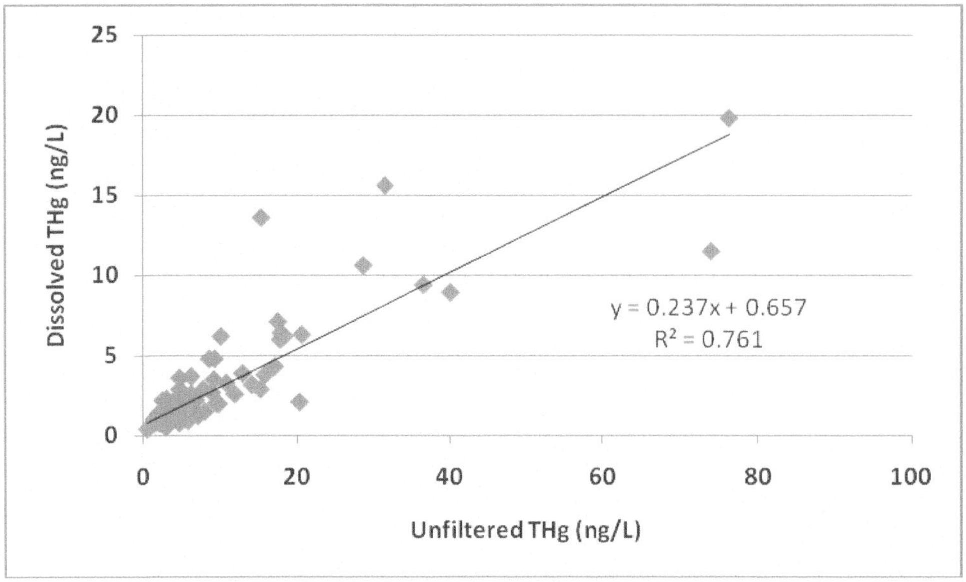

Figure A-1. Relationship between Dissolved (filtered) and Unfiltered T-Hg for Nine WWTPs. Data Source: NACWA 2007.

A.3.2 Partition Coefficients for Total Mercury

The measurement of dissolved Hg levels and TSS in the NACWA study also gave us an opportunity to calculate partition coefficients (log values) for T-Hg and expand the database for WWTP effluents. Calculated partition coefficients for the nine WWTPs at which dissolved T-Hg and TSS were measured in plant effluent ranged from 5.05 to 5.80, with an overall average of 5.51. Standard deviations for the log partition coefficient ranged from 0.17 to 0.42 and with an overall value of 0.35. There are 129 individual observations from the nine plants. Overall, these data are consistent with the data presented in Chapter 3.0. In looking at the data, it appeared that there was some relationship between the value of K_D and the TSS level in the effluent. In some cases, there was a strong negative relationship between these two variables at individual plants, but this was generally not the case. However, taking all the observations as a whole, there was a negative trend between TSS and K_D; that is, higher values of K_D were associated with lower TSS effluent values. Figure 6-2 shows the relationship between the partition coefficient and TSS for these plants. Note that the partition coefficient appears to be stronger at lower TSS levels. This type of relationship is frequently seen with mercury and other metals. One interpretation for this is that at lower TSS levels, the particles are smaller and more reactive, and therefore bind mercury more strongly. Another interpretation, as discussed in Chapter 3.0, is that there is an increase in the "colloidal" fraction which passes the filter with increasing TSS (i.e. these two variables are related) and this leads to an overall decrease in the measured KD with increasing TSS.

At some plants, there was also a relationship between T-Hg and the partition coefficient. However, the relationship was sometimes positive and sometimes negative. Considering all the data, the correlation coefficient between T-Hg and K_D was -0.10, which is indicative of no relationship. A relationship was noted between effluent T-Hg and TSS, although the NACWA report indicates that, considering all the data, there was little relationship between these two variables. It is of interest however, that the intercept of the T-Hg vs. TSS plot has an intercept of

2.8 ng/L. Hsu (2004) noted that the intercept for data plotted for other WWTPs was 5.8 ng/L. However, the slope of the relationship was is similar (0.524 – Hsu, 0.522 - NACWA).

Table A-2. T-Hg Partition Coefficients and their Relationship to TSS in WWTP Effluent. Data Source: NACWA 2007.

WWTP	Average Log K_D	Standard Deviation	Number of Observations	Average Effluent TSS (mg/L)	Correlation Coefficient (K_D:TSS)
A (secondary)	5.41	0.24	7	11.0	-0.08
B (activated sludge)	5.78	0.25	36	3.5	-0.61
C (trickling filter)	5.52	0.23	36	4.3	-0.32
G (tertiary)	5.24	0.42	12	4.7	-0.32
H (advanced)	5.80	0.23	3	2.9	-1.0
I (activated sludge secondary)	5.26	0.17	7	9.5	-0.50
J (secondary)	5.30	0.20	7	15.3	-0.82
K (secondary)	5.05	0.20	6	10.0	0.17
L (secondary)	5.42	0.42	15	8.9	-0.21
Overall	5.51	0.35	129	6.1	-0.47

A.4 Literature Review of Technologies for Mercury Removal from Wastewater Effluents

As described above, this section is divided into two parts. The first describes technology reviews for mercury removal from wastewater found in the literature that have been performed by others. These technologies tend to be more well-established. The second part reviews individual technologies identified as part of the current literature review. These technologies have been primarily investigated at the bench or pilot plant scale and are considered to be emerging technologies. The literature reviewed in this section dates from 1995 or later.

One issue with almost all the technologies reviewed in this section is that the investigations were conducted using industrial wastewater as opposed to municipal wastewater. Another issue is that the typical influent concentrations tested were much higher than those that typically occur in municipal wastewater. Therefore, the application of these test results to municipal wastewaters should be made with caution. It is entirely possible that technologies that give a >90% recovery at mg/L concentrations of Hg will not give a similar degree of efficiency when the concentrations are μg/L or ng/L. For instance, speciation of mercury in dilute solutions (e.g. complexation by DOC), unimportant at higher Hg concentrations, could have a considerable effect on the results at lower concentrations. For this reason, it is recommended that any promising technologies be tested on municipal wastewater effluents so that treatment efficiencies can be properly evaluated.

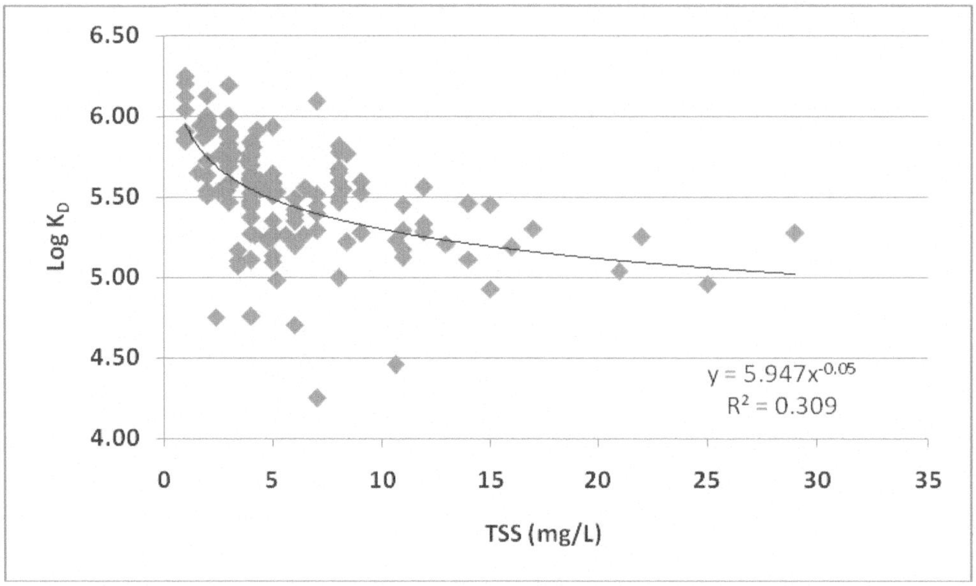

Figure A-2. Relationship between Partition Coefficient and TSS in Effluent for Nine WWTPs. Data Source: NACWA 2007.

It should also be noted that some very good work has been done by organizations like the American Dental Association (ADA) and the National Association of Clean Water Agencies on the evaluation of source control measures to reduce mercury in WWTP influent, especially in the area of dental amalgam (ADA 2007). This research effort is primarily focused on treatment rather than pretreatment and therefore source control efforts are not evaluated here.

A.4.1 Review Papers

A.4.1.1 U.S. EPA Capsule Report (1997)

U.S. EPA (1997) reviewed established and emerging treatment technologies. The established technologies consisted of:

- Precipitation
- Coagulation/co-precipitation, and
- Activated carbon adsorption.

Other less well-established and emerging technologies described included:

- Ion exchange,
- Chemical reduction,
- Membrane separation,
- Macrocycles adsorption,
- Biological treatment, and
- Membrane extraction.

The precipitation process described in the U.S. EPA Capsule report is the sulfide process. In this well-known process, a sulfide salt (e.g. sodium sulfide) is added to the wastewater.

Mercury precipitates out as mercuric sulfide particles[25]. Particulates are then separated from the bulk solution by gravity settling or filtration. Flocculants are sometimes used to enhance particle removal. Sulfide precipitation can achieve 99.9+ percent removal and has been commonly used in the chloralkali industry for wastewater treatment.

According to the capsule report, the most common coagulants used for Hg treatment in wastewater include alum, iron salts, and lime. The removal mechanism is through the incorporation of Hg ions into the matrix of the precipitated solid or adsorption to precipitated solids such as amorphous aluminum or iron hydroxides. This process is again followed by gravity settling or filtration to separate solids from the bulk solution. In studies of domestic sewage dosed with inorganic mercury followed by aluminum and iron co-precipitation, treatment rates were said to be on the order of 94-98%[26].

Adsorption by granular activated carbon (GAC) is another well-established means of removal of both organic and inorganic compounds from wastewater. In this process, wastewater is passed through packed beds of carbon and removal of mercury is achieved by sorption onto the solid matrix. Removal efficiencies reported in the Capsule Report ranged from about 50 to 99.9%. The report mentions that other sorbents have been investigated/utilized as well including starch xanthate, bicarbonate-treated peanut hull carbon (BPHC), modified *Hardwikia binata* bark (MHBB), coal fly-ash and the Forager sponge. In several bench-scale applications reported, starch xanthate adsorption resulted in 99.8% removal or better. Removal efficiencies were not reported for the remaining treatments.

According to the Capsule Report, ion exchange of Hg is limited to applications in which anionic mercury species are present, such as in chloralkali wastewaters where high chloride content results in negatively charged chloride species such as $HgCl_3^-$. Of course in domestic wastewater treatment plants, chloride concentration would tend to be low and DOC (or metal sulfide:DOC complexes) would likely dominate the speciation of Hg in solution. Therefore, it is unlikely that anion exchange would be employed for the treatment of domestic wastewater. A variation on ion exchange includes the use of chelating resins. In particular, the use of a weakly acidic polystyrene/divinylbenzene cation resin with thiol functional groups was reported to have been used successfully in a full-scale application at a defense installation.

The Capsule Report also touches on the use of chemical reductants. The concept is to reduce Hg^{II} to Hg^0. Since Hg^0 is volatile, it would be released in gaseous form or perhaps recovered as elemental mercury. Studies with reductants showed that this was a less efficient process and would likely require a polishing step to achieve lower levels. Membrane processes including reverse osmosis and micro- or ultra-filtration were discussed. These processes were reported to achieve removals ranging from 41 to 99.8% removals. Removal efficiencies were not

---

[25] The solubility of Hg in the presence of HgS solids is about 1 nM (200 ng/L) so it is unlikely that the solubility will be exceeded in many WWTPs. It is much more likely that there would be precipitation of FeS solids with the Hg being incorporated into these solids during precipitation or absorbed to the solid surface. This solubility is based on an 'estimated' intrinsic solubility constant - not one that was experimentally determined or listed in the NIST database. The estimate also assumes the presence of the aqueous HgS (or Hg(HS)OH) which is based on intrinsic ZnS and CdS mineral solubilities that were likely treating metal-sulfide nanoparticles as 'soluble' MeS(aq).

If HgS(aq) species are neglected, then cinnabar/metacinnabar are oversaturated when total [Hg]=0.01nM, S(-II)=1uM.

[26] Hg absorption onto oxide surfaces is well-known and is likely enhanced by the presence of DOC.

reported for other emerging technologies, with the exception of biological treatment, which was reported to remove between 7-99.5% of the influent mercury under a range of influent concentration, residence time and temperature conditions.

Table A-3 taken from U.S. EPA (1997) provides treatment efficiencies for the technologies reviewed in the report. Unfortunately, the best achievable result is about 0.5 ug/L (500 ng/L), which is far above the water quality targets established in most mercury TMDLs for the protection of human health or wildlife. In addition, a level of 500 ng/L is above the typical influent concentration in domestic sewage. Therefore, the technologies identified in this report would likely have to be enhanced in some way to make them applicable for domestic wastewater effluent applications.

U.S. EPA (2007) provides an update to its 1997 report. In this edition, the Agency reviews four categories of technologies for mercury treatment in water including precipitation/coprecipitation, adsorption, biological treatment and emerging technologies. Unfortunately, all of the case studies cited for performance and cost are for industrial or laboratory wastewater, or hazardous waste cleanup applications, and feature low-volume, high concentration (µg/L) levels and higher) waste streams. Therefore, none of the applications provide data of particular relevance to high volume, low concentration municipal wastewater situations.

Table A-3. Removal Efficiencies for Mercury Treatment Technologies. Source: U.S. EPA 1997.

Treatment Process	Achievable Concentration (µ/L)
Sulfide precipitation (+ filtration)	10 to 100
Co-precipitation	0.5 to 5
Activated carbon	0.5 to 20
Starch Xanthate	5 to 20
Ion Exchange	0.5 to 5.0
Reduction	10 to > 100
Membrane separation	*
* Membrane technology typically achieves 80-90% rejection, which is a measure of salt rejected into the brine.	

A.4.1.2 Department of Energy (Oak Ridge) Technology Review

About the same time as the U.S. EPA Capsule Report, the Department of Energy (Lanning 1993) released a review of aqueous treatment technologies relevant to mercury contamination in wastewater at the Oak Ridge Y-12 Plant in Oak Ridge, TN. In addition to the technologies reviewed in the Capsule Report, the DOE report mentions a sorption/filtration technology developed by the University of Washington which utilized a ferrihydrite coating for sand particles, presumably for sand filtration. Once the filter is loaded to capacity, sorbed metals are removed using a pH-induced desorption process. No additional performance details for this technology are provided.

Also mentioned in the DOE report is an ion exchange process which involves the preparation of a resin that bids both metallic cations and oxyanions, excluding chlorides or sulfides. AlgaSORB, a proprietary, nonliving, immobilized biomass developed by Bio-Recovery Systems, Inc. was identified as one such resin. Unfortunately, treatment was effective only to ug/L levels and further treatment would be required to meet today's more stringent regulatory levels. No further performance data was given for this technology.

The report also mentions an electrocoagulation process which is said to electrolytically treat wastewater resulting in the precipitation of suspended solids and certain metals and organics. Although no removal efficiencies are given for Hg, removal of As, Cd, and TSS were said to be 37%, 97-99.8%, and 94%, respectively, with capital and operating costs ranging from $0.25 to $0.50/1000 gallons.

Based on the results of literature review and vendor surveys, the DOE report recommended that filtration be evaluated in batch studies using standard GAC, sulfur-impregnated GAC, and a thio-group chelating ion exchange resin.

A.4.1.3 MASCO Feasibility Studies

In 1997, MASCO, Inc. (Medical Academic and Scientific Community) identified technologies for mercury removal from wastewater. Candidate technologies were:

- Simple Filtration
- Membrane Micro (and Nano) Filtration
- Reverse Osmosis
- Ion Exchange
- Chemical Precipitation/Redox Reactions
- Adsorption, and
- Disinfection

On the basis of their literature review MASCO identified 12 vendors who could provide technologies in four process groups, including activated/modified carbons, other specialized sorbents, enhanced filtration, and electrolytic precipitation. Vendors were interviewed and selected vendors were invited to participate in bench-scale feasibility testing of their technologies. The wastewater selected for treatment was from a local clinical hospital laboratory containing between 11 and 90 ug/L of total mercury. Testing was done on 5 gallon samples. Testing of the influent and effluent was performed using U.S. EPA Method 245.1. Results of the feasibility testing are show in Table A-4. Although some percent removals are in the high 90% range, the final effluent concentrations are typically not low enough for practical use in municipal wastewater treatment without further enhancement.

Table A-4. Summary of Wastewater Mercury Removals from MASCO Bench-scale Feasibility Testing Project. Data Source: MASCO (1997).

Participating Vendor (technology)	Number of Test Runs	Influent Mercury (µg/L)	Final Effluent Mercury (µg/L)	Test Removals (%)
ATA Technologies Corporation (chemically-enhanced sorbent)	1	33.0 - 41.3	0.112	99.7
Barnebey & Sutcliffe Corporation (activated carbon)	10	21.8 - 24.9	5.16 - 14.2	-
ICET, Inc. (activated carbon/modified filter media)	4	12.8 - 17.1	0.1 - 4.8	71.7 - 99.3
KDF Fluid Treatment, Inc. (galvanic cell)	2	33.0 - 41.3	18.4 - 20.2	45.6 - 50.5
Soils N.V. (electrolysis)	8	10.8 - 17.6	0.8 - 5.0	63.2 - 94.1
SolmeteX, Inc. (mercury selective sorbent)	9	12.8 - 24.9	0.114 - 1.1	94.4 - 99.2

A.4.1.4 California Central Valley Clean Water Agency Advanced Wastewater Treatment Capabilities: Evaluation of Research Needs

Most recently, the Central Valley Clean Water Association sponsored a study to evaluate research needs related to advanced wastewater treatment capabilities to meet the California Toxics Rule requirements (CTR) (WYA 2007). As part of the study a literature review was conducted to identify the published research that describes or demonstrates process performance for various treatment technologies and wastewater constituents. Mercury is one of the constituents covered by the CTR. Interestingly, this study, conducted a decade after the EPA Capsule Report, came to similar conclusions regarding the state-of-the-science with regard to treatment of mercury (and other metals); that is, that the existing literature does not provide an adequate basis to recommend advanced treatment technologies for treatment of mercury (and other metals). Specifically, the report found that very few studies provided enough information to allow for meaningful comparisons of results to regulatory standards and that no significant information was found regarding the effectiveness of secondary treatment for mercury.

As part of this study, data was also collected from 17 advanced wastewater treatment plants in California which employed either activated sludge with denitrification, granular and or cloth media filtration, and/or chlorine disinfection. Flows from these facilities ranged from 3 to 167 MGD. Results of the data analysis reveal that about 10% of the facilities discharged an effluent containing less than 1 ng/L T-Hg and about 93% of the facilities discharged less than 10 ng/L. Less than 1% of the plants discharged greater than 100 ng/L of T-Hg.

A.4.2 Emerging Technology Review

Although a number of papers were found dated prior to 1995, they are not reviewed in this section. The primary rationale for this decision is two-fold. First, prior to 1995 it is unlikely that clean techniques would have been used in sampling and Method 1631 was probably not used to testing. Therefore, studies that may have considered low mercury levels would be suspect. In addition, review papers written in the mid-1990s are likely to have adequately covered the existing technologies at that time.

The reviews performed as part of the current study are divided into physical, chemical, and biological technologies. Methods that involve the reduction of Hg^{II} to Hg^0 appear to be the most promising of those reviewed because they lead to a reduction of bioavailable mercury in the effluent. Reduction of Hg^{II} to Hg^0 provides the opportunity for volatilization of gaseous Hg to the atmosphere. However, a number of reduction-based technologies have implemented means of capturing and recovering the gas-phase mercury as part of the process. Further discussion is provided in Section A.5.

Most of the technologies reviewed in this section are summarized in the Table A-5.

A.4.2.1 Physical Technologies

Several papers were found that investigated the use of non-conventional sorbents such as carbonized coirpith (a waste solid generated when separating the fiber and husk of coconut, Namasivayam and Kadirvelu, 1999), steamed hoof powder (Ansari et al., 1999), acrylamide-amended coconut husk (Sreehar and Anirudhan, 2000), activated carbon derived from fertilizer waste (Mohan et al., 2000), carbonized Sugi wood (Pulido-Novicio et al., 2001), and banana stalk-polyacrylamide grafted copolymer (Shibi and Anirudhan, 2002). However, they were not reviewed here because of the presumed scarcity of the sorbent source in the U.S.

Navarro et al. (1996) studied the removal of mercury from wastewater using porous cellulose carrier modified with polyethyleneimine. Batch experiments demonstrated that the sorbent would selectively remove mercury even under acidic regions conditions. Sorbent capacity was calculated from measurements to be 288 mg/g (1.44 mM/g).

Feng et al. (1997) described the technology developed to produce Self-Assembled Mercaptan on Mesoporous Silica (SAMMS) for mercury removal from aqueous wastewater and a detailed paper on various SAMMS (Self-Assembled Monolayers on Mesoporous Supports) sorbents, their synthesis, and structure is provided by Fryxell et al. (2007). This technology has been licensed to a private company with reported success; however, documentation of these applications appears to be unavailable.

Hollerman et al. (1999) continued the earlier work at Oak Ridge National Laboratory (ORNL) by testing various sorbents of low-level mercury waste. The best of the sorbents tested had thiol-active sites. However, none of the sorbents tested reduced mercury concentrations to less than 12 ng/L. The average inlet mercury in these tests was 1093 ng/L. This testing led to the conclusion that, due to the large volume of sorbent required, a wastewater treatment plant using this technology would be prohibitively expensive to construct and operate. Other tests conducted using a Forager sponge and other sorbents (Taylor and Klasson, 1999) showed effective Hg removal, even at high flow rates. In these tests, Hg influent concentrations averaged about 275 ng/L. However, when mercury in the influent was associated with fine particles, making it less available to the sorbents, the removal efficiency dropped. Chlorination, which may have converted Hg to an ionic form, improved sorbent performance[27]. Sorbents were reported not to

[27] It is likely that the destruction of DOC by chlorination lead to higher removal efficiencies as the formerly DOC-bound would be released to solution where it would be available for adsorption by solids.

Table A-5. Summary of Emerging Technologies Reviewed for Removal of Mercury from Wastewater.

Technology	Hg Levels Tested	Wastewater Type	Result	Source
Physical				
Forager sponge sorbent	10 µg/L	Industrial wastewater, DOE weapons plant	> 99.9% removal	Taylor (2005)
Granular activated carbon (GAC)	3.8 mg/L	Pharmaceutical wastewater (66% organic Hg, Thimerosal)	98.8 to > 99.9% removal	Cyr et al. (2002)
Granular activated carbon (GAC)	100 to 700 mg/L	Laboratory simulated wastewater (mercuric chloride in double-distilled water)	91 to 100% removal Sorbent capacity 38.5 mg/g	Kannan and Rajakumar (2003)
Natural clays impregnated with 2-mercaptobenzimidazole	50 mg/L	Laboratory prepared solutions	> 99% removal Sorbent capacity 29 to 36 mg/g	Manohar et al. (2002)
Poly(2-hydrohyethylmaethacrylate) metal chelating gel beads modified with poly(ethyleneimine)	50 to 500 mg/L	Laboratory prepared solutions	Sorbent capacity 334 mg/g	Denizli et al. 2003)
Chemical				
Lithium-intercalated transition metal dichalcogenides	2000 mg/L	Laboratory prepared solutions	Sorbent capacity 580 mg/g	Gash et al. (1998)
Ion exchange	70 to 90 mg/L	Mine wastewater	> 99.9% removal Resin capacity 439 mg/g	Monteagudo and Ortiz (2000)
Ion exchange	10 mg/L	Chloralkali process wastewater	> 99.9% Resin capacity 712 mg/g	Dujardin et al. (2000)

Table A-5. Summary of Emerging Technologies Reviewed for Removal of Mercury from Wastewater (continued).

Technology	Hg Levels Tested	Wastewater Type	Result	Source
Semiconductor photocatalysis	120 mg/L	Laboratory prepared solutions	70%	Chen and Ray (2001)
$SnCl_2$ reduction followed by air stripping	138 ng/L	Groundwater	> 94%, varied with reagent dose	Looney et al. (2001 and 2003)
Complexation / ultrafiltration	5 to 10 mg/L	Laboratory prepared solutions	Approached 100% at pH > 7	Barron-Zambrano et al. (2002)
Amalgamation with copper shavings	1 to 10,000 ug/L	Contaminated groundwater	96 to 98% removal	Huttenloch et al. (2003)
Modified sulfide precipitation	20 to 30 ug/L	Chloralkali wastewater	99.9%	Tonini et al. (2003)
Biological				
Microbial reduction *Pseudomonas putida*	1.6 to 7.6 mg/L	Chloralkali wastewater	90 to 98% removal	von Canstein et al. (1999)
Microbial reduction *Pseudomonas putida*	40 mg/L	Laboratory prepared solutions	92 to 98% removal	Okino et al. (2000)
Microbial reduction *Pseudomonas*	3 to 10 mg/L	Neutralized chloralkali electrolysis wastewater	97% removal	Wagner-Doebler et al. (2000)
Microbial reduction *E. coli*	2.58 mg/L	Wastewater from a contaminated site	99% removal	Deng and Wilson (2003)
Microbial reduction *Acidithiobacillus ferrooxidans* SUG2-2	20 ng/L	Laboratory wastewater	47 to 100% depending upon destruction of organics in pretreatment	Takeuchi et al. (2003)
Sulfide precipitation in constructed wetlands	37 ng/L	Groundwater, stormwater	80%	Nelson et al. (2004)

demonstrate consistent mercury removal capabilities over time. More recently, Taylor (2005) reported that Forager M-TU (Dynaphore, Inc.) provided >99.9% removal of mercury from dilute wastewater in short term column tests with a contact time of 120 seconds. The nominal influentconcentration of Hg in these tests was 10 μg/L. ORNL operates a 0.5MGD wastewater treatment with a GAC packed columns designed for removal of non-volatile organics. These columns also remove about 75% of the mercury in the water. Additional mercury removal occurs in the filter columns prior to the GAC columns, since a variable portion of the mercury is attached to particulates. The GAC columns typically last at least five years, since the organic concentration in the wastewater is very low. Column changeout has not occurred for mercury loading, so ORNL does not have costs directly associated with mercury treatment (Taylor 2007, personal communication).

Cyr et al. (2002) reported on the results of a pilot scale GAC system to remove mercury from pharmaceutical wastewater. The sorbent used was F-400 (Calgon Carbon Co., Huntington, VA). Total mercury levels in the wastewater were quite high (3.8 mg/L) relative to raw municipal wastewater and a significant percentage of the total mercury, about 66%, was organic (Timerosal), with the remainder reported as inorganic mercury. TSS of the wastewater averaged 49 mg/L and the average pH was 8.3. The mercury concentration in the effluent from the pilot plant carbon columns (two columns in series) ranged between 0.5 and 37 ug/L over 157 hrs of operation, representing removal efficiencies of > 99.9% and 98.8%, respectively. The study concluded that removal of mercury to below the detection limit [of Method 245.1] of 0.2 ug/L may not be technically or economically feasible for pharmaceutical wastewater.

In a search for more cost effective sorbents, Manohar et al. (2002) turned to natural clays which had been impregnated with 2-mercaptobenzimidazole. Removal of Hg^{II} was found to be > 99% at an initial concentration of 50 mg/L. Adsorption increased with increasing pH and was optimal in the range of 4 to 8. Mercury uptake by the clay was also found to increase with the ionic strength, temperature, sorbent dose, and smaller sorbent particle size. The clay was > 90% kaolinite. Adorption maxima ranged between 29 to 36 mg/g (0.14 to 0.18 mM/g).

In addition to studying indigenously prepared carbons from wood and cashew nut shell, Kannan and Rajakumar (2003) also studied the removal of mercury from aqueous media using commercial activated carbon. As the tests of Manohar also demonstrated, the authors showed that the percentage removal of Hg increased with sorbent dose, pH, temperature, ionic strength and lower initial concentration of Hg ions and sorbent particle size. Aqueous mercury concentrations ranged from 100 to 700 mg/L in these tests. At 35 degrees-C, Hg removal at 100 mg/L was 100% and decreased monotonically to 91.1% at 700 mg/L. With Hg concentration in the hundreds of mg/L range, maximum adsorption was calculated to be 38.5 mg/g (0.19 mM/g).

Denizli et al. (2003) studied the removal of aqueous mercury from synthetic solutions using poly(2-hydrohyethylmaethacrylate) gel beads modified with poly(ethyleneimine). These metal-chelating beads sorbed mercury quickly (equilibrium was reached within 1 hr). Maximum adsorption of 1.67 mM/g (occurred at pH 5. Regeneration was performed with 0.1 M HNO_3.

A.4.2.2 Chemical Technologies

A small number of studies were found which investigated chemical technologies for mercury treatment.

Lithium-intercalated transition metal dichalcogenides, Li_xES_2 ($0.25 < x < 1.9$; E = Mo, W, Ti, Ta), were studied by Gash et al. (1998) for the extraction of aqueous metal ions including Hg^{2+}. For Li_xTiS_2 and Li_xTaS_2, hydrolysis produced S^{2-} ions which precipitated Hg^{II} as HgS. Li_xMoS_2 did not undergo hydrolysis, however. When isolated following exchange and heated to 425 degrees-C, Hg_yMoS_2 released nearly a quantitative amount of mercury and produced essentially mercury free MoS_2. Samples of MoS_2 were reactivated by treatment with n-butyllithium with no loss of exchange capacity. The ratio of the volume of recovered mercury (secondary waste) to the volume of 10 mM (2000 mg/L) aqueous primary waste was 1.5E-4.

Monteagudo and Ortiz (2000) investigated the use of three ion exchange resins for treatment of mercury in mine wastewater. Dowex XZS-1, a strong cationic ion exchange gel had the greatest selectivity for mercury with a Langmuir adsorption maximum of 4.39 meq/g (439 mg/g or 2.19 mM/g). Using this resin influent levels ranging from 70 to 90 mg/L were treated to 34 ug/L (> 99.9% removal).

Dujardin et al. (2000) used ion-exchange resins bearing thiol groups to remove mercury from chloralkali process wastewater. A macroporous styrene-divinylbenzene copolymer (TMR, Rohm and Haas), had a Langmuir saturation parameter calculated to be 3.56 mM/g (712 mg/g). Using this resin, wastewater initially containing 10 mg/L-Hg was reduced to 1 ug/L, a removal efficiency of > 99.9%.

Removal of mercury from wastewater was studied using semiconductor photocatalysis by Chen and Ray (2001). This relatively new technique couples low-energy ultraviolet light with semiconductor particles acting as a catalyst and is based on metal reduction by the photo-generated electrons. It was observed that photoreduced Hg (along with silver) was deposited on the catalyst while other metal ions remained in solution. Two TiO_2 catalysts (Degussa P25, Degussa Company and Hombikat UV100, Sachtleben Chemie GmbH) were investigated in this study. Introduced reductants (ferrous iron and EDTA) improved the photoreduction of mercury while higher levels of dissolved oxygen hindered it. Initial Hg concentration in solution was 0.6 mM (120 mg/L). Removals on the order of 70% were observed when the solutions were purged with nitrogen. Removal efficiencies dropped to about 30% when the test solution was purged with air and about 15% when the test solution was purged with pure oxygen[28].

Looney et al. (2001 and 2003) investigated the removal of mercury from low concentration aqueous streams using chemical reductants and air stripping. The process consists of dosing the water with low levels of stannous chloride (Sn^{II}) to convert the mercury to Hg^0. This mercury species can then be easily removed from the water by air stripping. Groundwater containing mercury concentrations of approximately 138 ng/L (0.69 nM) was used for the study. In undosed samples, sparging removed none of the initial mercury. Removal in the treated samples varied by reagent dose. Low doses (<1 nM stannous chloride) showed little removal. Stannous chloride doses above 50 nM showed relatively complete removal (>94%) and final mercury concentrations < 10 ng/L (<0.05 nM). At stannous chloride doses between 1 and 50 nM, mercury removal was a function of the dose. In general, effective mercury removal was achieved with doses ranging from approximately 5x to 25x the theoretical minimum stoichiometry (based on the redox reaction between Hg^{II} and Sn^{II}).

Bhandare and Arkegar (2002) investigated the carrier-facilitated transport of Hg^{II} against its concentration gradient from aqueous 0.1 M HCl solution across a flat-sheet supported liquid membrane (SLM) containing triisobutylphopshine sulfide (Cyanex 471X) as the liquid carrier in kerosene. Dilute sodium thiocyanate solution (0.11 M) was the most efficient stripping agent among several investigated. Under optimum conditions, the transport of Hg^{II} across the liquid

[28] The use of nitrogen would enhance reduction of Hg while the presence of oxygen would enhance oxidation and therefore reduce net reduction.

membrane was about 100% after 6 hrs. This method has been demonstrated to selectively recover mercury from waste samples and mercurochrome solutions.

Polyethyleneimine (PEI) was used as a complexing agent by Barron-Zambrano et al. (2002) to investigate the removal of mercury from aqueous solutions by complexation-ultrafiltration. This method uses water soluble polymers to bind metals ions from aqueous solution followed by ultrafiltration to remove the metal-polymer complexes from solution. An advantage of the method is said to be the low energy requirement. Initial Hg concentration in solution was 10 mg/L. Mercury retention by complexation/ultrafiltration strongly depends upon the pH and chloride concentration in solution. At pH 1, retention is near zero even with no chlorides present. However, at a chloride concentration of 1M, zero retention occurred up to pH 5. Mercury retention approaching 100% can be achieved for pH > 7.

Huttenloch et al. (2003) investigated the use of copper shavings to remove mercury from contaminated groundwater or wastewater by amalgamation. Amalgamation/sorption results were expressed in terms of the Freundlich isotherm. Retention on copper shavings ranged from 16.22 to 0.018 mM/g for commercial shavings and from 12.48 to 0.030 mM/g for recycled shavings, with the lower results obtained from Hg in 0.01M $CaCl_2$ solution and the higher results from Hg in distilled water. The concentration range studies was 1 to 10,000 ug/L. Batch sorption experiments showed that 96-98% of the Hg in solution was removed within 2 hr. Removal of mobilized Cu^{2+} ions resulting from amalgamation by some means (the authors suggest an ion exchanger such as zeolites) would likely be necessary in order to realistically implement this process.

Tonini et al. (2003) reported on the modification of an existing sulfide precipitation process at a chloralkali facility to achieve low levels of mercury in discharged effluent. Typical removal efficiencies for the raw wastewater were 99.9% resulting in effluent concentrations of 20-30 ug/L. The primary modification to the system involved the use of MERSORB LW, a sulfur impregnated activated carbon, downstream of the sulfide precipitation/filtration processes. Initial testing required modifications to the treatment train to achieve reasonable results. These changes included the addition of ferric chloride to the effluent of the sulfide treatment process (involving another filtration step) and the adjustment of the influent pH to about 3.5. These changes, together with the addition of adsorption, were able to consistently treat effluent waters to < 100 ng/L with some results approaching 20 ng/L (average result was 66 ng/L over the observational period). When groundwater was introduced into the system on the shutdown of the mercury cell manufacturing operations, a silica removal (precipitation) pretreatment system was also required.

A.4.2.3 Biological Technologies

In 1999, von Canstein et al., reported the removal of mercury from chloralkali electrolysis wastewater by a mercury resistant *Pseudomonas putida* bacterial strain, isolated from polluted river sediments. The bacterium is able to reduce ionic to metallic mercury. Effluents from several chloralkali plants in Europe, ranging from 1.6 to 7.6 mg/L, high chloride concentrations (up to 25 g/L) and widely varying pH (2.4 to 13) were tested by passing them through biofilms of *P. putida* Spi3 grown on a porous media carrier. Mercury retention by this system was shown to be between 90 and 98%.

About the same time, Okino et al. (2000) reported the development of a biological mercury removal-recovery system using the same bacterium. Optimum temperature, pH, thiol concentrations and cell concerntrations for removal of mercuric chloride were determined. From 92 to 98% of the initial 40 mg/L of Hg was recovered in 24 hr. Recovery rates of Hg from river and seawater were reported to be of the same order as from the laboratory solution in experiments with the same initial concentration.

Similarly, during the same time period, Filali et al. (2000) identified several strains of bacteria that were resistant to heavy metal toxicity, among them *Klebsella pneumoniae*, in sewage water of Casablanca, Morocco. These bacteria were identified as being resistant to mercury, copper, and zinc but were suspectible to cadmium and silver toxicity. These bacteria were also shown to be resistant to antibiotics and aromatic hydrocarbons. Later, Zeroual (2001) reported that immobilized *K. pneumoniae* cells isolated from wastewater effectively reduce and volatilize Hg in a fluidized bead reactor. The highest cleanup efficiency and volatilization rates were obtained when *K. pneumonia*e was entrapped in alginate beads.

In 2000, Wagner-Doebler et al., reported the removal of mercury from chemical wastewater by microorganisms on a technical scale. Pure cultures of seven mercury resistant strains of *Pseudomonas* were immobilized on a carrier material inside a 700L packed bed bioreactor. Neutralized chloralkali electrolysis wastewater having an Hg concentration of between 3 and 10 mg/L was continuously fed through the bioreactor at flow rates from 0.7 m^3/hr up to 1.2 m^3/hr. Mercury retention efficiency was 97% within 10 hrs of inoculation and effluent concentratrions were consistently below 50 ug/L. Filtration by activated carbon after the bioreactor further reduced the effluent to below 10 ug/L. The efficiency of the bioreactor was unaffected by conductivity variations between 20 and 105 uS/cm and pH variations from 6.5 to 7.5. Temperature increase up to 47°C did not impair performance of the bioreactor and standby periods of up to 6 hr were well-tolerated. Wagner-Doebler et al. (2000a) showed the biofilms used in their bioreactor to consist of a monolayer of bacteria. Droplets several microns in diameter which accumulated outside the bacterial cells were identified as elemental mercury by electron-dispersive X-ray analysis. The initially present monospecies biofilm of *Pseudomonas putida* Spi3 was invaded during the course of the experiment by 13 other strains. Of these, only two were not mercury resistant, while all others had resistance levels similar to or higher than *P. putida*. Further studies by von Canstein et al. (2001) reported that the bioreactor showed no signs of depleted performance after 16 months of operation. Von Canstein (2002) also showed that bacterial species diversity improved performance of packed-bed reactors. Mixed culture biofilms displayed a higher tolerance for rapid increases in or continually high Hg concentrations. Species diversity decreased with stress induced by higher Hg concentrations but was completely restored after concentrations returned to normal. Wagner-Doebler (2003) reported packed-bed reactor biotechnology to be a cost-effective alternative to other current treatment technologies. Deckwer et al. (2004) extended the work of this group to include the microbial reduction technology in a three-phase fluidized bed (TPFB) reactor process. The first phase involves mercury reduction on an alginate immobilized biocatalyst, the second the removal of Hg^0 into the gas phase, and the third, the recovery from the gas phase by fast oxidative absorption. Although the biotransformation properties of this process are similar to the fixed bed technology, the authors claim that the TPFB may be more reliable and less costly than the former.Another bacterium, *Escherichia coli*, was genetically engineered by Bae et al. (2001) for the enhanced uptake and bioaccumulation of mercury. Synthetic phytochelatins (ECs) are a class of metal-binding peptides with a repetitive binding motif, Glu-Cys$)_n$Gly, reported to bind heavy metals more effectively than metallothioneins. However, limited uptake across the cellular memebrane is often the rate-limiting factor for intercellular bioaccumulation by organisms expressing these peptides. Two strategies were successful in enhancing transport across the cell membrane and increasing bioaccumulation of Hg^{2+}; either coexpressing an Hg^{2+} transport system with (Glu-Cys)$_{20}$Gly(E20) or by directly expressing EC20 on the cell surface.

Deng and Wilson (2003) demonstrated the ability of genetically engineered *E-coli* to remove mercury from wastewater. A hollow-fiber bioreactor was used to retain induced cells for continuous mercury uptake. The cells removed more than 99% of the mercury in the wastewater (2.58 mg/L, pH 9.6) and the final amount of Hg accumulated was 26.8 mg/g (0.13 mM/g), while none of the other ions were removed from the aqueous media. EDTA also stimulated Hg uptake.

Takeuchi et al. (2003) showed that *Acidithiobacillus ferrooxidans* SUG2-2 cells were also capable of reducing mercury from Hg^{2+} to Hg^0 under acidic conditions. The authors found that mercury reduction and volatilization was strongly inhibited by $AgNO_3$, K_2CrO_7, cysteine, trichloroethylene and commercially produced detergents. However, the inhibition was not observed when the organic compounds were chemically decomposed using Fenton's method before bacterial treatment. Pretreatment to decompose organic compounds improved the reduction and volatiliazation of Hg from 47% to nearly 100%.

In addition to the applications described involving bacterial reduction of mercury, another class of "biotechnology" exists in the use of constructed wetlands for mercury removal from aqueous media. In this case, an anaerobic environment is created in the constructed wetlands. This reducing environment promotes sulfate reduction, producing sulfides, which precipitate mercury and other divalent cations. This technology has been demonstrated at the Department of Energy's Savannah River site (King et al. 2001). Removal rates have been reported by Nelson et al. (2004) in excess of 80%. Potential issues with this technology include the potential production of methylmercury (King et al. 2001). The authors postulated that the formation of dissolved sulfide complexes may be responsible for the increases in MeHg production. Nelson et al. (2004) also noted that the wetland was a net source of DOC, which may have mitigating effect on the bioavailability of dissolved inorganic mercury. Although effluent total mercury concentrations were not given, MeHg levels at 0.26 ng/L were said to be 3.5% of the discharged T-Hg suggesting an effluent concentration of 7.4 ng/L. Furthermore, 80% removal suggests influent levels of about 37 ng/L.

A.5 Cost of Mercury Removal

The costs of mercury removal, either by existing or emerging technologies, are not well-established. In a report for the Clean Estuary Partnership (CEP, 2003), unit costs for filtration, reverse osmosis, and filtration plus RO for San Francisco Bay Area WWTPs were derived from cost estimates contained in 1993 National Research Council publication titled *Managing Wastewater in Coastal Urban Areas* (NRC, 1993). Annual unit costs (expressed as $ million per year per MGD) were derived from the information provided in the NRC publication:

- Filtration 0.18-0.65
- Filtration plus RO 1.20

These costs are derived from annualized capital and annual operation and maintenance costs and are indexed to a 2001 construction cost index of 6342. The source document for these costs included costs with an estimated 1991 construction cost index of 4835.

The estimated mercury load reductions resulting from the additional treatment options described above are based on assumed removal efficiencies. For filtration, the assumed mercury removal efficiency was calculated from the median effluent concentration values for estimated secondary (13.6 ng/L) versus advanced treatment (4.8 ng/L). Mercury removal efficiencies for reverse osmosis were calculated assuming a final RO effluent concentration of 2 ng/L. Final effluent concentrations were calculated for each facility based on the following assumed treatment efficiencies:

- Filtration 65 % removal of Hg from secondary effluent
- Reverse Osmosis 60 % removal of Hg from filtered secondary effluent

In the 1997 Capsule Report, U.S. EPA presents a cost comparison for xanthate treatment and sulfide precipitation treatment for 50 m^3/day (~12,500 gal/day) contaminated with 3 mg/L Hg. An investment capital cost of $280,000 (1987 dollars) was assumed for both plants. Operating costs were not greatly different; $77,500 for xanthate treatment versus $91,000 for sulfide precipitation.

Lanning (1993) provided a cost estimate to treat a 20 gpm stream (0.029 MGD) from 70 ug/L to 2 ug/L (97% removal) of $14,000 for two staged RO units with operating costs of $1.00/1000 gallons (about $1670/lb of mercury).

Cyr et al. (2002) performed a cost analysis of the removal of mercury from pharmaceutical wastewater using GAC. At a flow rate of 1 gal/min (equivalent to 1440 gal/day) the capital cost of the pilot system was $0.012/gal and the operating cost was $0.095/gal, a total of about $0.11/gal or $110/1000 gal (about $4,030/lb of mercury removed).

Wagner-Doebler (2003) provided a cost analysis and comparison for the treatment of 8 m^3/day (~2,000 gal/day) of chloralkali wastewater at an initial concentration of 4 mg/L using regenerable ion exchange columns and biological mercury retention. Capital costs (in euros) were 1.3 MM for the ion exchange columns and about 0.5 MM for the microbial technology. Annual operating costs for the microbial technology (90,000) were estimated to be about 37% of the ion exchange costs (240,000). Overall, the cost of removal per kg was 760 euros for the microbial technology versus 1,400 euros per kg for ion exchange. Removal cost per on a USD/lb basis would be about $484/lb for the microbial technology and about $890/lb for ion exchange. Although Deckwer et al. (2004) do not provide a cost comparison of their TPFB reactor to the packed bed technology compared to ion exchange above, the authors suggest that the former would be cheaper to operate because the TPFB does not require regeneration of the biocatalyst.

Cost estimates for the technologies investigated in the MASCO were prepared by most of the vendors. The technologies included GAC filtration, electrolysis, and proprietary sorbents. Cost estimates for these technologies are presented in Table A-6 along with those for other emerging technologies. Cost estimates were standardized to the degree possible. Since the most commonly estimated flow rate was 2,000 gal/day, the cost estimate provided for this flow rate was provided in the table where possible. Cost estimates made in different years were brought to 2007 dollars by using a simple inflation (3%) rate factor [$f = 1 + 0.03(n)$] where n is the number of years from the year of the estimate to 2007. Values presented in Euros were converted to US dollars (USD) by multiplying by 1.4 USD/Euro. All capital costs were annualized using an 8% interest rate and assuming a 10 year project life. Annualized total costs were estimated by adding the annual operating costs (2007$) to the annualized capital cost (2007$).

Table A-6 shows a wide range of annualized costs for the various technologies, almost two orders of magnitude, from ~$7,250/yr for filtration using a proprietary sorbent to

~$680,100/yr for ion exchange. The RO costs presented by Lanning (1993) seem low, as is apparent when the annualized treatment costs are normalized to the flow rate. By comparison to the costs from the CEP report (0.18 to 1.2 $/yr/gpd), these costs are quite high. Recall, however, that the CEP cost estimates were for large plants treating millions of gallons per day while the costs provided in Table A-6 are for much smaller systems. Economies of scale would likely reduce annual cost if these technologies were to be implemented at the flow rates of a typical WWTP.

Table A-6. Comparison of Annualized Costs for Various Mercury Removal Technologies.

Technology	Flow Rate (gpd)	Estimate Date	Capital Cost ($)	Operating Cost ($/yr)	Capital Cost (2007 $)	Operating Cost (2007 $)	Annualized Total Cost (2007$/yr)	Annualized Unit Cost ($/yr/gpd)	Reference
Proprietary Adsorbent	2,000	1997	35,000	360	45,500	468	7,249	3.624	MASCO (1997)
GAC	1,440	2002	7,400	6,188	8,510	7,116	8,384	5.823	Cyr et al. (2002)
GAC	2,000	1997	3,800	12,614	4,940	16,398	17,134	8.567	MASCO (1997)
Reverse Osmosis	28,800	1993	14,000	10,512	19,880	14,927	17,890	0.621	Lanning (1993)
Bipolar Electrolysis	2,000	1997	61,000	24,090	79,300	31,317	43,135	21.568	MASCO (1997)
Proprietary Adsorbent	2,000	1997	39,000	73,000	50,700	94,900	102,456	51.228	MASCO (1997)
Proprietary Adsorbent	2,000	1997	150,000	100,000	195,000	130,000	159,061	79.530	MASCO (1997)
Xanthate Adsorbent	12,500	1987	280,000	77,500	448,000	124,000	190,765	15.261	US EPA (1997)
Sulfide Precipitation	12,500	1987	280,000	91,000	448,000	145,600	212,365	16.989	US EPA (1997)
Microbial Reduction	2,000	2003	700,000	126,000	784,000	141,120	257,959	128.980	Wagner-Doebler (2003)
Ion Exchange	2,000	2000	1,820,000	336,000	2,038,400	376,320	680,102	340.051	Wagner-Doebler (2003)

Discount rate (i) = 0.08
Project Life (yrs) n = 10
2007 $ were computed using a simple interest 3% inflation rate for the number of years since the date of the estimate.

A.6 Effects of Wastewater Treatment on the Bioavailable Fraction of Mercury

This section will introduce a new concept related to the enhancement or mitigation of bioavailability; that is the effect of wastewater treatment processes themselves on the bioavailable fraction of Hg in the effluent. It is important to remember in considering this concept that although a certain process may actually increase the bioavailable fraction of mercury in the effluent, overall WWTPs typically provide greater than 90% reduction of mercury across the treatment plant. Thus, even though the bioavailable fraction may increase, the mass of Hg available for bioaccumulation or transformation to bioavailable species is greatly reduced by treatment. In addition, it should be recalled from Chapter 3.0 that WWTPs appear to have a singular ability to produce Hg-sulfide complexes that, even though they are measured in the dissolved fraction, may not be bioavailable and appear particularly resistant to degradation.

In performing this assessment, it was essential to identify wastewater treatment processes commonly used in the United States. The researchers used the 10 treatment trains identified by the National Research Council (NRC, 1993). For each of these treatment trains (collections of unit processes), they postulated whether the treatment involved would tend to increase or decrease the bioavailable fraction, based on the description provided by the NRC of the treatment system. The following is reprinted with permission from the National Academies Press. Copyright 1993, National Academy of Sciences.

> *"A large number of technically feasible wastewater treatment technologies are currently available. Ten representative systems, arranged roughly from the simplest to the most complex, have been selected to demonstrate the wide range of treatment*

capabilities and costs. The ten systems are all proven technologies in full-scale operation in the United States. The data presented below for these ten systems can be used to make comparative judgments regarding performance, to estimate the approximate costs of meeting various effluent discharge standards, and to compare the costs of point and nonpoint source treatment options.

The ten wastewater treatment systems are as follows:

1. Primary (PRI)

2. Chemically enhanced primary (CEPT)

a. low-dose chemically-enhanced primary (CEPT)

b. high-dose chemically-enhanced primary (HD-CEPT)

3. Conventional primary + biological treatment (BIO)

4. Chemically-enhanced primary + biological treatment (CEPT-BIO)

5. Primary or chemically enhanced primary + nutrient removal (NUTR)

6. System 5 + gravity filtration (NUTR-FILT)

7. System 5 + high lime + filtration (NUTR-HI-FILT)

8. System 5 + granular activated carbon + filtration (NUTR-FILT-GAC)

9. System 5 + high lime + filtration + granular activated carbon (NUTRHI-FILT-GAC)

10. System 9 + reverse osmosis (NUTR-HI-FILT-GAC-RO).

Common elements for all treatment systems presented in this report include influent pumping, preliminary treatment (bar screens and grit removal), effluent disinfection (chlorination and dechlorination), effluent pumping, and a sludge processing system consisting of dissolved air flotation of biological sludges, and anaerobic digestion of combined primary and thickened biological sludges, followed by belt press dewatering.

Description of Ten Wastewater Treatment Systems

1. Primary Treatment

Primary treatment is a physical process that involves gravity separation of settleable and floatable solids from the influent wastewater stream. Removal of settleable solids takes out some associated pollutants, including organic matter, nutrients, heavy metals, toxic organics, and pathogens. Other physical separation processes, such as fine screens and filters can be included in this treatment step.

2. Chemically-Enhanced Primary Treatment

Chemically enhanced primary treatment is a modification of the primary clarification process through the use of chemical coagulants, typically metal salts and/or organic polyelectrolytes. By varying the chemical dose, the performance of chemically enhanced clarification systems can be adjusted to increase the removal of suspended solids, BOD, and/or total phosphorus. Chemically-enhanced primary treatment facilities can be divided into two categories:

a. Low-dose chemically-enhanced primary treatment is used mainly for increasing the removal of suspended solids, BOD, metals, and toxic organics. A low-dose chemically-enhanced primary treatment plant is defined as the addition of a metal salt or other primary coagulant in concentrations between 5 mg/l and 100 mg/l, with or without the application of a polymer, prior to primary clarification.

b. High-dose chemically-enhanced primary treatment is used mainly to increase the removal of suspended solids, BOD, metals, and toxic organics in addition to the removal of phosphorus. The added metal salts react with soluble ortho-phosphate in the influent wastewater, producing a precipitate that is removed in the waste sludge. High-dose chemically-enhanced treatment is defined as the addition of a metal salt or other primary coagulant in concentrations greater than 100 mg/l, with or without the application of a polymer, prior to primary clarification.

Chemically enhanced primary treatment provides opportunities for size reduction of follow-on treatment; if iron salts are used, the control of hydrogen sulfide, a major cause of odor problems; and in some cases, the potential for production of extra methane as a fuel source.

3. Conventional Primary + Biological Treatment

Conventional biological treatment systems, often classified as either suspended (e.g., activated sludge) or attached growth systems (e.g., trickling filters), use a diverse culture of microorganisms to break down organic matter in the wastewater, oxidizing a portion and converting the remainder into biological solids. Organic contaminants are removed by biodegradation and volatilization. Nondegradable suspended contaminants are removed by physical entrapment and subsequent removal with the generated biomass. Some soluble constituents (i.e., heavy metals) are removed by adsorption on the biomass. Some nutrient removal occurs through incorporation into the generated biomass. Biological treatment systems convert some influent organic nitrogen and urea to ammonia, thereby increasing the ammonia concentration making it more biologically available upon effluent discharge. Some biological systems are operated to convert ammonia to nitrate. Gas-liquid mass transfer is required to supply oxygen to the biological process, and this often results in stripping and gaseous discharge of volatile compounds. With disinfection, this effluent is sometimes used for irrigation of agricultural lands. In fact, effluent from all of the following treatment trains can be reclaimed for certain uses.

4. Chemically-Enhanced Primary + Biological Treatment

Chemically enhanced primary + biological treatment involves the use of metal salts and polymers with either a conventional or innovative biological treatment system. Chemically-enhanced primary treatment has three major effects on a biological system-enhanced removal of phosphorus (HDCEPT), potential improved BOD removal by the biological system due to enhanced efficiency of the primary treatment stage, and increased sludge quantity. Raw sludge production is increased in the chemically-enhanced primary stage and decreased in the biological stage such that the overall amount of raw sludge is greater by 10% to 20% compared with a conventional primary +

biological treatment system. However, there is less than a 10% difference in the overall amount of digested sludge produced by the two systems due to the destruction of the volatile content in the chemically enhanced primary stage (Chaudhary et al., 1991).

5. Nutrient Removal

Wastewater treatment systems can be configured to remove the nutrients nitrogen and/or phosphorus. Nitrogen removal is accomplished by an extension of the conventional biological system to incorporate the biochemical processes of nitrification and denitrification. Nitrification is the oxidation of ammonia and organic nitrogen to nitrate nitrogen. The process is mediated by the activity of a specialized class of autotrophic bacteria that can be grown in conventional activated sludge biological systems by extending the biological solids residence time resulting in more complete biodegradation of organic matter. Nitrogen removal is subsequently obtained by denitrification whereby the nitrate nitrogen is reduced to nitrogen gas and then released into the atmosphere. Phosphorus removal can be accomplished by chemical or biological means. High-dose metal salts addition, as described in the section on chemically enhanced treatment, results in phosphorus removal. Alternatively, biological phosphorus removal can be accomplished through the selection of high phosphorus content microorganisms, resulting in a greater mass of phosphorus in the excess biological solids removed. Biological phosphorus removal systems are more capital cost-intensive and less operations and maintenance cost-intensive than chemical phosphorus removal systems and their efficiency can vary depending on a number of factors. Consequently, biological phosphorus removal systems typically incorporate some degree of chemical addition (usually for polishing) to ensure reliability and low phosphorus concentrations in the effluent.

6. Nutrient Removal with Gravity Filtration

This alternative includes a filtration system in addition to the nutrient removal system. This combination will remove additional quantities of TSS, along with other contaminants associated with the TSS (such as BOD5, nitrogen, phosphorus, and heavy metals). The capability to add chemicals to the effluent filters is also provided, allowing further removal of phosphorus and other pollutants. Sludge production is increased slightly with this alternative relative to the nutrient removal system alone due to the increased removal of pollutants. Effluent filtration will increase the removal of pathogenic organisms, metals, and toxics from the treated effluent and enhance the performance of downstream disinfection processes. This system is used in some areas to produce water for use in urban irrigation.

7. Nutrient Removal with High Lime

This alternative follows a nutrient removal system with a high lime treatment system. High lime treatment involves the addition of lime (CaO) to elevate the pH of the treated effluent to over 11.0. This process provides considerable removal of phosphorus and heavy metals, removal of high molecular weight residual organics, and disinfection of the treated effluent. Solids produced as a result of lime addition are separated in a downstream clarifier and the pH of the clarifier effluent is adjusted back to neutral using either carbon dioxide or acid. Alternatively, two pH adjustment steps (referred to as two-stage recarbonation) can be used to remove excess calcium and magnesium from the treated effluent. In the first step, the pH of the clarified effluent is reduced to 9.3 to allow precipitation of calcium carbonate, which is removed in a downstream clarifier. The pH of the effluent from this second clarifier is subsequently reduced to 7.0. Because a high level of phosphorus removal will be obtained independent of influent phosphorus concentration, upstream removal of phosphorus by either biological or chemical means is not necessary. Large quantities of chemical sludge are produced in the high lime treatment process. The chemical sludge can either be dewatered and land-filled or it can be processed to thermally regenerate lime, suitable for reuse in the same process. In either case, processing of the lime sludge is relatively difficult and expensive. For the purposes of this analysis, thermal regeneration of the lime is assumed.

8. Nutrient Removal and Granular Activated Carbon

This alternative adds the granular activated carbon (GAC) system to the NUTR-FILT system. GAC is a physical process for the removal of residual organic materials, including toxic organics, from a treated wastewater effluent. Some heavy metals may be removed as well. Treated effluent is applied to downflow packed beds containing granular activated carbon. Residual organic materials are removed as the treated effluent passes through the packed beds. GAC has a fixed capacity for removal of organics and, when this capacity is fully utilized, the beds must be removed from service and the exhausted GAC regenerated. Thermal regeneration is typically used and is assumed for this case.

9. Nutrient Removal with High Lime and Granular Activated Carbon

This alternative, the first of two potable reuse alternatives described, adds high lime and GAC to the nutrient removal system. This process with disinfection has typically been applied to wastewater that is reclaimed for indirect potable reuse.

10. Nutrient Removal with High Lime, Granular Activated Carbon, and Reverse Osmosis

This alternative, the second of two potable reuse alternatives described, involves an additional treatment step, a reverse osmosis (RO) system, after System 9. The RO process is applied typically when dissolved solids removal is required to prevent salt build-up within a recycling system. RO involves application of highly treated effluent under high pressure to a membrane. It allows the water to flow through but is not permeable to (i.e., rejects) dissolved solids. Dissolved solids are concentrated as a brine in a reject stream. RO's best application is as a complementary process with GAC to accomplish complete removals of a broad spectrum of pollutants. Operating costs for the RO process are high due to the energy costs of maintaining high pressure, limited membrane life due to fouling, and the high cost of brine disposal. Owing to its high cost, RO has been applied primarily for water reuse applications in areas where water is scarce and expensive. In many instances, only a portion of the treated effluent is processed through RO."

Table A-7 shows the results of the assessment of the effect of these wastewater treatment processes on the Hg bioavailable fraction[29]. Also included in the table are typical T-Hg removal efficiencies[30], effluent concentrations, percent dissolved Hg, and percent MeHg, where they could be estimated from data or literature sources.

While wastewater treatment plants are particularly effective at reducing total mercury, the nature of the processes tends to increase the percentage of dissolved mercury in the effluent compared to the influent. As shown in Table A-5, primary treatment alone removes on the order of 80% of the mercury, while secondary and advanced treatment removes from 90 to over 99%. The fraction of dissolved mercury in the effluent appears to range from around 50% for secondary treatment to around 70% for advanced treatment. The average dissolved: total fraction of mercury in the WWTPs in Table 3-1, for which there are data, is about 70%. By the research team's definition, this shift to a greater fraction of dissolved mercury would indicate an increase in the bioavilable fraction and therefore bioavailability. Likewise, any production of MeHg or increase in the percentage MeHg would be defined as an increase in the bioavailable fraction. The caution is that these data do not reveal the percentage of the dissolved Hg that is complexed by organic or inorganic ligands, which would tend to mitigate increases in bioavailability. Therefore, it is probably necessary to also consider the impact of treatment on the DOC (or NOM) levels in the WWTP waters. If these increase across the plant, then this may mitigate the relative increase in dissolved Hg concentration across the plant.

Filtration processes would tend to increase the bioavailable fraction as they remove particulate mercury (which by the team's definition is not bioavailable) while having little or no effect on dissolved mercury, thereby increasing the percentage of dissolved Hg. Thus, primary treatment (Configuation 1; PRI) is assumed to increase the fraction that is bioavailable.

Configurations 2a and 2b are thought to primarily increase bioavailability due to settling of particulate mercury. The addition of metal salts or organic polyelectrolytes as coagulants would tend to enhance settling and therefore increase bioavailability. On the other hand, there might be a reaction between the added coagulant and mercury in solution. If this occurred, it might reduce ionic mercury in solution, resulting in a decrease of bioavailability. Overall, however, the effect of primary treatment, whether it is enhanced by the addition of coagulants or not, is probably to increase bioavailability of Hg in the effluent.

---

[29] The researchers re-emphasize that most of the treatments remove Hg relatively effectively, primarily through the removal of solids-associated Hg. Thus, in terms of bioavailability, as defined, solids removal will lead to an increase in the fraction of the dissolved Hg, and thus the bioavailable fraction, even though the total Hg concentration is being decreased dramatically. In addition, potential increase in bioavailability would be mitigated to an extent by the fraction of inorganic dissolved Hg complexed by DOM and DOM-S linkages. It is not clear in each instance if the outflow from each treatment would therefore have a lower concentration of bioavailable Hg. In the discussion below, the effect of each treatment is discussed in terms of the bioavailable fraction as this does not require knowledge of T-Hg, but the relationship between the two definitions of bioavailable Hg should be kept in mind. In addition, the conclusions reached pertain primarily to inorganic mercury, for several reasons. First, the fraction of MeHg is small, usually on the order of a few percent. Second, they have defined both particulate and dissolved MeHg to be bioavailable. Therefore, the removal of any form of MeHg would result in a decrease in Hg bioavailable fraction and bioavailable Hg.

[30] The percent removals in Table A-7 are cumulative for the listed process and other stages leading up to it.

Table A-7. Observed Removal Efficiencies, Effluent Characteristics, and Effect on Mercury Bioavailable Fraction of Conventional and Advanced Wastewater Treatment Technologies.

Treatment Configuration	T-Hg Removal (%)	T-Hg Effluent Concentration (ng/L)	Dissolved T-Hg (%)	MeHg (%)	Effect of Treatment on Bioavailable Fraction	Plant Location	Reference	Comments
1. Primary (PRI)	80 43	 49.3	 5.2	 2.6	Increase	St. Paul, MN Sacramento, CA	Balogh and Liang (1995) Parmer et al. (2005)	
2a. Low-dose CEPT					Primarily increase			Decrease in bioavailability may result from reaction of ionic Hg with added metal salt
2b. High-dose CEPT					Same as 2a.			Same as 2a.
3. PRI + biological treatment (BIO)	90 – 97 99 (est.) 97.6 95-96 94	3 – 15 4 – 25 (with recycled scrubber incinerator water) 2.3 31 2.6 4.92	 56% 54% 30-40% 28%	 2.2 11 7.7	Primarily increase	Winnipeg, Manitoba Cleveland, OH Athens, GA Duluth, MN Various Sacramento, CA	Bodaly et al. (1998) NEORSD (2006) U.S. EPA (2000) WLSSD (2007) NACWA (2007) Parmer et al. (2005)	Some decrease in bioavailability may result from the adsorption and removal of ionic mercury with the biological waste solids
4. CEPT + BIO					Same as 3.			Same as 3.
5. PRI or CEPT + nutrient removal (NUTR)					Primarily decrease			Increase in bioavailability may result if MeHg is produced during anaerobic treatment
6. System 5 + gravity filtration (FILT)	99 98	2	70 60	2	Primarily increase	San Jose, CA	TetraTech (2007) NACWA (2007)	
7. System 5 + high lime + FILT					Decrease			
8. System 5 + granular activated carbon (GAC) + FILT		10% < 1 93% < 10 99% < 100			Increase	Central Valley, CA	WYA (2007)	
9. System 5 + high lime + FILT + GAC					Additional filtration would tend to increase bioavailability			
10. System 9 + reverse osmosis					Decrease			

WERF

Configurations 3 and 4 essentially reflect secondary treatment, which probably typical of most U.S. treatment plants. These configurations encompass trickling filters as well as activated sludge systems. It is assumed that solids removal following biological treatment is accomplished with either clarifiers or filtration. Bioavailability is assumed to be primarily enhanced by this process. This is logical because it is a process in which organic matter is being broken down by microbes for energy. In this process, Hg would be released from the organic matter. Because it is an aerobic process, mercury would tend to be converted to or remain in an ionic form (Hg^{II}). Most of this ionic mercury will be incorporated into biological solids via adsorption, although some may be incorporated into the biomass itself. Because most of the mercury entering the biological treatment phase of the treatment system would be retained in the secondary sludge, this process is assumed to increase the bioavailability of the mercury remaining in the effluent.

Configurations 5 through 10 utilize biological nutrient removal (BNR). Both the removal of N and P in BNR processes involve the creation of anaerobic conditions. As such, reduction of mercury to Hg^0 may occur, either due to the presence of reducing conditions or via the action of anaerobic bacteria. Volatilization of reduced Hg could occur along with the evolution of nitrogen gas during denitrification. Ionic mercury may also be taken up by bacteria during P removal. As such, the BNR process is assumed to decrease the bioavailability of Hg in the effluent, since it would tend to reduce dissolved ionic Hg. However, it is possible that in the presence of sulfates under reduced conditions, MeHg might be formed through the activity of sulfate reducing bacteria. In most of the treatment plants investigated in this study, however, the formation of methylmercury seems to be very limited. In the New York treatment plants shown previously in Table 3-1, the MeHg content of the effluent ranges very narrowly from 1.1 to 3.1% of T-Hg. In the effluent of the North Oconee treatment plant (Athens, GA) the MeHg/T-Hg ratio was 2.2%. Likewise, in the San Jose/Santa Clara, CA plant, the ratio was 2%. However, in two plants for which data was accessible (SCRSD and WLSSD) the ratios are higher in the 9-11% range[31].

Configuration 6 adds gravity filtration following BNR and configuration 8 adds granular activated carbon (GAC). As stated above, it was assumed that removal of solids by filtration increases bioavailability. Even though GAC filtration would tend to remove some colloidal material, like DOC, Hg complexed by DOC is by definition not bioavailable and the weakly complexed ionic Hg that passes through the GAC is; therefore the researchers made the assumption that filtration by GAC would tend to increase bioavilability.

Configurations 7 and 9 involve the addition of lime following BNR. The addition of lime would increase pH and tend to promote the formation of hydroxides and carbonates of mercury. In this way, the concentration of ionic mercury would be reduced, and the research team assumes that the process would result in a net decrease in bioavailability.

Configuration 10, involves the addition of reverse osmosis, which would tend to lower the ionic strength of the effluent overall, probably resulting in lower concentrations as ionic

[31] It would be remiss to mention that the ratio observed for the WLSSD plant is based on only three observations. It is also the case for this plant that the influent background is at about 6% MeHg (n=4). Therefore, the contribution from the plant may be only about 5%, which still appears to be greater than most other plants for which data is available. WLSSD operates an anaerobic sludge digestion process which recycles sludge decant water back to the headworks of the plant. The sludge decant water from this anaerobic process may be a possible source of elevated MeHg in the effluent.

mercury in solution. For this reason, the team assumes that the use of RO would result in a net decrease in bioavailability.

Table A-8 summarizes the effects of the emerging technologies reviewed in this section on Hg bioavailable fraction. These technologies have been arranged in the table from lowest to highest anticipated cost and within each cost range (L, M, H) by increasing removal effectiveness. In terms of reducing bioavailability of effluents, the most promising technologies would be those that have higher Hg removal, lower cost, and that *decrease* the Hg bioavailable fraction. In these terms, chemical or biological reduction would appear to be the most promising technology because of their intermediate cost, relatively high removal efficiency, and anticipated reduction of bioavailability[32]. Although it is not listed in the table, disinfection may also have an impact on bioavailability. It should be noted that the anticipated effect of disinfection on the bioavailable fraction is an increase, due to the fact that destruction of organic matter would likely release Hg^{II} into solution.

Another technology to be considered is the oxidation of strongly complexed Hg that may be coming from the anaerobic processes within the treatment plant. To the extent that these complexes enhance bioavailability fraction, oxidation may result in adsorption to particulates that may be subsequently settled out prior to discharge. On the other hand, if these complexes prove to mitigate bioavailability, there may be opportunities in the treatment process to promote their formation. Finally, trials are underway to evaluate coagulation of colloidal DOC bound mercury to further enhance removal. While this may be effective at removing additional mercury, the process would tend, by the research team's definition, to increase the bioavailable fraction of an effluent by decreasing mercury that is considered not be bioavailable, and thereby increasing bioavailability in the effluent. Better removal effectiveness and cost information will emerge as such studies are performed on low level mercury in municipal effluents. It should be emphasized that while some of these processes may tend to increase the bioavailable fraction, all the processes listed should dramatically reduce the bioavailable concentration of mercury in the effluent, as indicated by the high removal efficiencies for each.

Efforts to remove additional mercury must be evaluated in view of the effects that any such adjustments would have on other effluent characteristics, such as BOD, nutrients, etc. This underscores the need to evaluate any proposed changes to a treatment system at any given plant on a case-by-case basis. Generalizations for "tweaking" plant processes to achieve greater Hg removals at such low levels would be imprudent without proper study at an individual plant.

[32] At this time the researchers anticipate that chemical reduction would be more cost effective than biological reduction, as the one cost estimate the researchers have for biological reduction is quite high. However, this cost estimate was for removal of Hg from an influent stream with much higher concentrations and it may be that biological reduction of lower concentrations would be more cost-effective. Contributing to the lower cost of chemical reduction is that air sparging is relatively inexpensive. This cost could be higher if Hg^0 recovery were to be required.

Table A-8. Factors Affecting for Inorganic Mercury Removal and Mercury Bioavailable Fraction by Emerging Wastewater Treatment Technologies.

Treatment Technology	Removal Efficiency (%)	Cost	Effect on Hg Bioavailable Fraction	Comments
Filtration	65	L	Increase	Increase bioavailable fraction by filtering out Hg attached to solids, increasing the proportion of dissolved Hg
Treatment Wetlands	80	L	Unknown	May decrease bioavailable fraction by creation of insoluble Hg sulfides; may increase through creation of soluble Hg sulfides; may decrease by complexation with DOC; may increase by creation of MeHg
Absorption/Adsorption	70-99.9	L	Process dependent	Decrease Hg in solution if ionic forms are sorbed, increase if complexed forms are sorbed
Electrolysis	60-94	M	Decrease	Decrease ionic mercury by precipitation as an insoluble hydroxide or carbonate
Chemical/Biological/ Photolytic Reduction	70-97	M	Decrease	Decrease ionic Hg in solution by creating Hg^0
Reverse Osmosis	60	H	Decrease	Decrease ionic Hg in solution
Ion Exchange/Amalgamation	96-99.9+	H	Decrease	Decrease ionic Hg via exchange with other cations

APPENDIX B

MERCURY TMDL ALLOCATIONS

B.1 Introduction

Emission of mercury, in either particulate or gaseous forms, and its subsequent deposition to landscapes and waterbodies, is a major concern of the U.S. EPA for mercury under the Clean Water Act (CWA). In rivers, lakes and estuaries, atmospherically deposited divalent (inorganic) mercury, either deposited directly to the lake surface or transported from the watershed, may become transformed into methylmercury, a species more toxic and bioaccumulative than inorganic mercury, enhancing its entry into aquatic food chains. Input of MeHg from the atmosphere is relatively small in most situations. Other point and non-point sources, not directly related to deposition, may also contribute inorganic Hg and MeHg to waterbodies. As methyl mercury works its way up the food chain, it becomes present at increasingly higher levels in fish tissue and at increasingly greater percentages of the total mercury present. According to U.S. EPA, mercury levels present in fish tissue above 0.3 mg kg^{-1} are of concern to the general population of human consumers.

Many states have issued fish consumption advisories restricting fish consumption based on U.S. EPA's reference dose and levels observed in fish tissue. Many of these same states have adopted the position that restriction of fishing and/or fish consumption in these waterbodies constitutes an impairment of the resource and have placed them on the state's list of impaired waters under CWA Section 303(d). Over two thousand waterbodies across the United States have been thus classified. In response to these listings and recent court orders, U.S. EPA and the states have developed and continue to develop mercury Total Maximum Daily Loads (TMDLs), a calculation of the maximum amount of mercury a waterbody can assimilate without exceeding the applicable water quality standard(s) for its designated use.

Wastewater treatment plants discharge mercury in wastewater directly to waterbodies and are regulated by the National Pollutant Discharge Elimination System (NPDES). Conventional wastewater treatment, sludge incineration and even sludge dewatering/drying processes may contribute quantities of mercury to the atmosphere as well as volatile losses may occur during these processes.

B.2 Background

B.2.1 Water Quality Criteria for Mercury

The primary objective of this appendix is to review mercury load reduction allocations that have been given to wastewater treatment plants as a result of the establishment of TMDLs. However, any discussion of TMDL allocations would be incomplete without an accompanying discussion of water quality targets (WQTs) which are used in TMDLs to establish overall reductions. In this regard, the wastewater treatment community faces an evolving regulatory landscape.

U.S. EPA's first recommendation of a Section 304(a) water quality criterion for mercury dates to 1976. In 1985, the water quality standard recommended by the U.S. EPA for the protection of aquatic life was 12 ng L^{-1} and this standard was adopted by many of the states. In 1995, the U.S. EPA revised its 304(a) acute criterion to 1.7 ug L^{-1} and the chronic criterion was increased from 0.012 ug L^{-1} to 0.91 ug L^{-1}. Meanwhile, the US Food and Drug Administration (FDA) had recommended fish tissue guidance of 1 mg kg^{-1} for protection of human health. A number of states adopted this guidance for the purpose of issuing fish consumption advisories. Other states have used a risk-based approach to set thresholds for fish consumption advisories. In many cases, the states have also used these thresholds as the targets in their TMDL analyses. As the reader will note in the following TMDL reviews, these values typically range from about 0.2 mg kg^{-1} to 0.4 mg kg^{-1}, depending upon the specific assumptions each state makes concerning parameters in the risk equations.

In January, 2001, the U.S. EPA recommended a human health criterion for mercury that was the first of its kind; a water quality criterion measured in fish tissue rather than in the water column. While many agree that this is a reasonable approach for mercury and other bioaccumulative substances, the Agency has left the states with little guidance as to how this criterion should be implemented in the regulatory process. This in turn has created a situation in which the use of the criterion for 303(d) listing, TMDLs, and NPDES permitting by U.S. EPA regions, the states, and potentially tribes, varies widely.

A few states have adopted a tissue based standard but only two have adopted 0.3 mg kg^{-1} as their water quality standard for mercury for the protection of human health. Others have adopted the *approach* U.S. EPA used to derive the criterion but have adapted it for local conditions by changing parameters such as fish consumption rates, relative source contribution, etc. with resulting values for tissue concentration ranging from 0.2 to 12 mg kg^{-1}. Still others retain water column based standards with values ranging from 1.8 ng L^{-1} to 600 ng L^{-1}. Many states have also interpreted their narrative toxics standard by placing waterbodies on their 303(d) list when fish consumption advisories are posted. Thus, the wastewater treatment community is faced with confusing array of standards across the nation.

Most states and tribes will likely move towards adopting U.S. EPA's criterion (or an allied criterion adjusted by State specific preferences) and the older criterion of 12 ng L^{-1} will fall into disuse. The 12 ng L^{-1} standard was confusing at best. Even though it was recommended as an aquatic life protection criterion, it was derived as a human health criterion (using the FDA guidance of 1 mg kg^{-1} and dividing by a bioconcentration factor). Many states had adopted it as "protective of all uses" since the acute and chronic criteria for protection of aquatic life were much higher. Many states appear to be waiting for U.S. EPA's final guidance on implementation of the methylmercury criterion to be released to adopt a tissue-based standard.

In addition, water quality targets for the protection on wildlife have also been developed and in some mercury TMDLs have the effect of lowering the target below what a value that might be derived for human health protection. In 1995 (U.S. EPA, 1995), the U.S. EPA derived a wildlife criterion that was assumed to be protective of both avian and mammalian fish consuming species, the lowest computed criterion being associated with the avians. This led to the adoption of a T-Hg criterion for the protection of wildlife in the Great Lakes of 1.3 ng/L. The use of wildlife criterion as TMDL targets is less common than a human health criterion; however, wildlife-driven TMDL targets have been increasingly utilized in recent TMDLs and appear to be gaining in popularity. One reason for this is the involvement of the U.S. Fish and

Wildlife Service in review of mercury TMDLs, especially where endangered species are deemed to be at risk. Given debate over the appropriateness of sublethal endpoints and the uncertainty in the data used to estimate effects levels, there is a risk that wildlife-based criteria will continue to go lower and continue to drive water quality and/or fish tissue targets down. Some TMDLs previously based on human health protection might also be reevaluated using wildlife criteria.

B.2.2 Review of Mercury TMDLs

The review of mercury TMDLs was performed with two objectives:

- to assess the types of allocations wastewater treatment plants receive in TMDLs which may ultimately be used to assess the level of the risk mercury TMDLs pose to the wastewater treatment community, and
- to help develop a basis for providing guidance to individual wastewater treatment plants that may be faced with mercury TMDLs.

In terms of the first objective, the reviews focused on the allocations developed for WWTPs in each specific TMDL. In terms of the second objective, the reviews focused on the approaches used by the U.S. EPA regions and states in various mercury TMDLs. These reviews therefore provide an opportunity to assess the defensibility of existing mercury TMDLs and provide a basis to develop guidance for individual facilities if they become involved in such a TMDL.

Nine recent TMDLs were reviewed as part of this project. In addition, the conclusions reached herein rely heavily on the earlier review of 73 TMDLs performed as part of a project for the Electric Power Research Institute (EPRI, 2006)[33]. Because many of these TMDLs have been done by specific states and U.S. EPA regions, the reviews are grouped along geographic lines in the following discussion. It will also be evident to the reader that the development of strategies for coping with mercury TMDLs at individual facilities must be tailored to the specific approaches used by the regions or states that have developed them, as each regulatory agency thus far has taken somewhat unique approaches.

In reviewing the TMDLs, instances were noted when specific wastewater treatment plants were identified. In some cases, mercury data from these facilities were summarized or available in the TMDL.

B.3 Key Findings

The following is a summary of the key findings from the TMDL reviews. It is based upon review of the 73 TMDLs as described in EPRI (2006) and the more recent TMDLs reviewed for this project. Since the review of more recent TMDLs has added little in the way of new information, these findings are taken largely from the EPRI report. Findings are summarized in four categories:

- Use of Fish Tissue Criterion as a Water Quality Standard;
- Water Column Targets Derived from Fish Tissue Targets;

[33] The number of TMDLs does not correspond to the number of listed waterbodies. In most instances, a single TMDL is done for all the listed segments in a waterbody or watershed. Therefore the TMDLs reviewed represent a much greater number of listed segments of waterbodies.

- Approaches for Deriving Water Column-Based Criteria; and
- Approaches to TMDL Allocations and Implementation.

B.3.1 Use of Fish Tissue Criterion as a Water Quality Standard

Regardless of whether U.S. EPA's tissue-based criterion has been formally adopted by the states, the criterion is used liberally in the establishment of TMDLs by the states and U.S. EPA regions. It is rare, in the TMDLs reviewed for this project, that the state or region uses the 0.3 mg kg^{-1} criterion without some modification of the assumptions used to derive it. The TMDLs done in the State of Georgia by U.S. EPA Region 4 are among the few in which the 0.3 mg kg^{-1} criterion is used without modification.

In some cases, the criterion is adjusted for local fish consumption rates. A case in point is the TMDL for San Francisco Bay, wherein the tissue-based target is calculated to be 0.2 mg kg^{-1}, based on a fish consumption rate of 32 g day^{-1} rather than the 17.5 g day^{-1} used by U.S. EPA to derive the 0.3 mg kg^{-1} criterion. The criterion is not often adjusted for the relative source contribution (RSC) although this has been done in some TMDLs (e.g. Minnesota). In some cases the tissue-based criterion is above the 0.3 mg kg^{-1} criterion, as in the Cashie River, NC, where the target is 0.4 mg kg^{-1} and is based on the criterion the State used to list the waterbody.

In a number of cases, the state or Region makes additional conservative assumptions with regard to the use of the tissue-based criterion. Although U.S. EPA derived the criterion using mean tissue concentrations in trophic levels 2, 3, and 4, and consumption-based weights for each trophic level, the criterion is often compared to the tissue concentrations in trophic level four fish only. The criterion is often compared specifically to largemouth bass, northern pike, walleye, or other top-level predator in the system tending to have the highest observed tissue levels. This adds an additional, but normally unquantified, margin of safety into the TMDL calculations.

In yet other cases, the states or regions standardize the criterion to a fish species of a certain length or age-class. For instance, the criterion may be stated as 0.4 mg kg^{-1} in a three-year old largemouth bass or 0.2 mg kg^{-1} in a 40 cm walleye. This is helpful in that it guarantees that the oldest, largest specimen with the highest observed tissue level is not used in the analysis.

While the states and tribes are free to develop their own criteria, alternative criteria must be based on scientifically defensible methods. It is not clear how U.S. EPA makes this determination as part of the TMDL review process.

B.3.2 Water Column Targets Derived from Fish Tissue Targets

In a number of TMDLs, the state or U.S. EPA Region derives a water column-based target from a fish tissue target. This results in a criterion for either total mercury or methyl mercury in the water column. Table 3-1 gives a summary of the derived criteria by U.S. EPA Region.

Inspection of the table shows that the results vary widely across the country, depending not only of the aquatic ecosystems involved but also the assumptions used by the agency establishing the TMDL. Total mercury values range from a low of 0.52 ng L^{-1} in the northern forests and wetlands of Minnesota to over 10 ng L^{-1} in California, which in some areas is impacted by mining activities[34]. Methylmercury criteria range from a low of 0.06 ng L^{-1} in

[34] These California rivers and streams tend to have low methyl to total mercury ratios.

mining impacted systems in California to 0.46 ng L^{-1} (nearly two orders of magnitude difference) in a blackwater Coastal Plain stream of North Carolina Coastal Plain. If these water column concentrations are derived from the fish tissue criterion using local empirical data, this suggests that for these systems that the bioaccumulation factors vary by at least an order of magnitude. If this is so, then this information argues strongly against using the default BAF values suggested by U.S. EPA.

B.3.3 Approaches for Deriving Water Column-Based Criteria

Water-column based criteria are derived in several ways from the tissue-based criterion, as follows:

- division of the tissue-based concentration by a bioaccumulation factor (BAF). In most cases these are calculated from measurements made in the waterbody. In limited instances, the TMDL uses default U.S. EPA BAFs[35],
- regression of waterbody-specific fish tissue concentrations on methylmercury concentrations in water, and
- regression of regional fish tissue concentrations on methylmercury concentrations in water.

B.3.4 Approaches for TMDL Allocations and Implementation

Calculation and allocation of load reductions and implementation of the TMDL in NPDES permits is of great importance to dischargers. In the TMDLs reviewed, calculation of load reductions has been accomplished in a variety of ways involving comparison of fish tissue, water column concentrations and sediment concentrations to targets developed for these media:

- Fish tissue concentrations are often compared directly to the fish tissue target (e.g. Region 6 TMDLs). As mentioned above, there are different ways the TMDLs make this comparison that affect the required load reductions.
- In some cases, ambient water quality concentrations are compared to the derived water quality targets (e.g. Willamette River TMDL, Maryland lake TMDLs, Sacramento-SanJoaquin Delta TMDL)[36]
- In the Georgia TMDLs, modeled water quality concentrations were compared to the water quality target.
- In the San Francisco Bay TMDL, current sediment concentrations were compared to a target level of 0.2 mg kg^{-1}.

In all cases, load reductions are required in the same percentage as the reductions required by comparison to fish tissue or water quality concentrations to the targets.

---

[35] As noted above, this approach is likely to have the largest associated error.

[36] The criteria and measured values may be in terms of total or methylmercury.

Table B-1. Summary of Fish Tissue and Water Quality Targets Calculated in Mercury TMDLs Nationwide.
Adapted with permission from EPRI (2006).[37]

Region (State – Type)	Total Mercury (ng/L)	Equivalent T-Hg Fish Tissue Conc. (mg/kg)	Methylmercury (ng/L)	Equivalent MeHg Fish Tissue Conc. (mg/kg)
EPA Region 3 (MD – Lakes)	1.32 [38]	0.235	-	-
EPA Region 4 (GA – Rivers)	3.8 [39]	-	-	-
EPA Region 4 (GA – Lakes)	7.0 [40]	-	-	-
EPA Region 4 (NC – River)	1.5 to 3.9 [41]	0.4	0.464 [42]	-
EPA Region 5 (MI – River)	1.3 [43]	-	-	-
EPA Region 5 (MN – Statewide)	0.52	0.2	-	-
EPA Region 9 (CA – Cache Creek & tributaries) Sacramento-San Joaquin Delta	13 - 41[44]	-	0.06 to 0.14 [45] 0.06	0.12 [46] 0.23 [47] 0.28 [48]
EPA Region 10 (OR – Willamette River	0.92	0.3	0.029 [49]	
Range	**0.52 - 41**	**0.2 to 0.4**	**0.029 to 0.46**	**0.12 to 0.28**

[37] Most of the water quality targets shown in the table are established by the TMDLs. In some cases, (e.g. Region 5, MN, the criteria established in the TMDL was the fish tissue target and the WQT was calculated in the TMDL for comparative purposes only. In some other cases, the WQTs were calculated by the authors from information in the TMDLs for comparative purposes.

[38] Average of 9 lakes.

[39] Average of 21 rivers.

[40] Average of 6 lakes.

[41] Estimated for this report based on the range of ratios of total methyl mercury to total mercury given in the TMDL.

[42] Total methyl mercury.

[43] Great Lakes Initiative (GLI) value for wildlife protection.

[44] Estimated for this report using ratios of total methyl mercury to total mercury given in the TMDL.

[45] Total methyl mercury.

[46] Trophic level 3 fish standard

[47] Trophic level 4 fish standard

[48] Standard for a 350 mm reference largemouth bass

[49] Dissolved methyl mercury

Once the load reductions are estimated, allocations are assigned to point and nonpoint loading sources to the watershed. These load allocations are referred to as the Waste Load Allocation (WLA) for point sources and the Load Allocation (LA) for nonpoint sources. In most cases, a Margin of Safety (MOS) is also included on the TMDL, which further restricts the WLA and LA[50]. Of particular importance to this review is the way in which allocations are made to the point sources. In the TMDLs reviewed, these allocations have been made in the following ways:

- In the event that a point source is categorized as a "major" source, it may receive an individual WLA. The definition of a major source varies from state to state[51].
- In TMDLs where point sources taken together are considered *de minimis*, they may receive an aggregated WLA[52].
- In TMDLs where the point sources taken together are considered *de minimis*, they may receive no WLA with language that there is to be no net increase in the discharged load (i.e. mercury discharges are capped at current levels).
- In some TMDLs point sources, such as WWTPs, have been assigned a zero WLA. This is particularly problematic since all point sources will discharge mercury at some level.
- In some TMDLs, wasteload allocations are deferred until monitoring results can be used to assign a WLA.

Implementation plans are not a required TMDL element. However, almost all TMDLs have some language with implementation recommendations for the permitting authority:

- While the language varies from state to state, in almost all cases point sources are required to monitor for mercury using an approved low-level method (i.e., Method 1631 for total mercury or Method 1630 for methylmercury).
- In almost all cases where a water quality target is assigned, point sources are required to monitor influent and effluent mercury and assess whether the facility discharge levels are above or below the WQT and whether the facility is adding mercury in its wastewater. This is almost always done on a concentration basis. Point sources whose effluents exceed the WQT and are adding mercury to the discharge are often required to develop plans of study to identify sources and mercury to their effluents and develop minimization plans to control those sources.
- In a number of cases where the state water quality standard is 12 ng L^{-1} and the point sources are considered *de minimis*, the TMDL has recommended no action with regard to changes in permits.

---

[50] Note that the U.S. EPA water quality fish criteria already has a safety factor (factor of 10) incorporated into its calculation.

[51] In the State of Georgia, for instance, a major source is one which discharges greater than 1 MGD. In North Carolina, a major source is any WWTP with a flow of greater than 1 MGD or that has a pretreatment program.

[52] The definition of a *de minimis* source may not be consistent among EPA regions. These particular allocation examples are taken from TMDLs done in EPA Region 4.

- In states such as Michigan and Ohio (MDEQ, 2004; Ohio EPA, 2000) where general permit variances have been established, permittees are given discharge limits based on "achievable levels" and develop characterization and minimization plans as a condition of the variance. In these states, the general variance appears to have been implemented in lieu of performing TMDL analyses (at least to date).
- In some states, guidance has been formalized for assessment of "reasonable potential (RP)"[53]. In Michigan, a statistical procedure is outlined to determine RP for facility effluents and in the draft Idaho guidance (IDEQ, 2004), the procedure for finding RP is based on fish tissue data exceeding a threshold in the waterbody to which the permittee discharges.

B.4 Review of Additional TMDLs

This section provides reviews of recent individual TMDLs reviewed for this project grouped by U.S. EPA Region, which were not reviewed in the prior EPRI report. Within each region, the individual states are arranged alphabetically. Within each state, TMDLs are also arranged alphabetically, providing the reader with an easy means of locating TMDLs that have been done in their U.S. EPA Region or state of interest. No new mercury TMDLs (since December 2005) were identified in regions 1, 2, 3, or 5.

B.4.1 U.S. EPA Region 4

Pascagoula River (MSDEQ, 2004) – This is a Phase 1 TMDL. The allocation for the three WWTPs is the permitted flow rate times the water quality standard of 12 ng/L, likely giving these facilities a target they can meet. However, there does not appear to be any data for the effluents of the point sources themselves. Facilities are being required to monitor for T-Hg using Method 1631. If T-Hg is detected (which is almost certain), then the facility may have to develop a minimization program.

B.4.2 U.S. EPA Region 6

Spring Lake AR (FTN, 2007) – There were no point sources in this watershed and the WLA was set to zero. Load reductions were calculated based on the ratio of T-Hg in a single largemouth bass fish tissue sample from 1992 and Arkansas' fish tissue target of 1 mg/kg. Including a 10% explicit margin of safety, the required load reduction from the watershed was estimated to be 14% of the current loading.

B.4.3 U.S. EPA Region 8

Ninemile Creek, MT – Although a reference to this TMDL was found, the TMDL itself could not be located. From the information found on the Internet, the water quality standard for T-Hg in Montana appears to be 50 ng/L. T-Hg loads were assigned under high and low flow regimes for the listed tributary creeks. No point sources were involved in the mercury TMDL; all load reductions were associated with sediment load reductions and restoration of mining sites.

---

[53] Reasonable potential is language adopted in the Clean Water Act meaning that a discharger is found to have reasonable potential to exceed a water quality standard based on its discharge characteristics.

B.4.4 U.S. EPA Region 9

Alamo Lake, AZ (ADEQ and Tetra-Tech, 2005) – This TMDL was focused on nonpoint sources; atmospheric deposition, mining, and natural areas. No point sources are located in the watershed and an allocation of zero was therefore assigned to point sources. The load reduction calculation was somewhat different from previous TMDLs done in Arizona and is worth mentioning. The overall reduction was estimated by calculating an upper confidence level (at the 95% confidence level) UCL_{95} on the observed mean of the largemouth bass fish tissue data. This value was then compared to the fish tissue target of 0.3 mg/kg. By this means, a 63% reduction in water column methylmercury was deduced to achieve the fish tissue standard. Among the margins of safety in this approach are that fish tissue concentrations are measured in T-Hg while the fish tissue target is assumed to be in MeHg, the UCL_{95} is used rather than the mean in calculating the required load reduction, and a trophic level four fish is used to compare to the 0.3 mg.kg standard rather than an average fish tissue residue value over all trophic levels.

Guadalupe River, CA (CRWQCB, 2006) – The Guadalupe River TMDL concludes that the major sources of mercury in the watershed are mining sites and that methylmercury observed in fish is primarily produced in reservoirs under low DO conditions. The problem is proposed to be solved by implementing “low-tech” erosion control of the sources. Other identified sources are atmospheric deposition, urban runoff, and naturally occurring mercury in soils.

Fish tissue targets were proposed for protection of wildlife (0.05 mg/kg in 50 to 150 mm fish and 0.10 mg/kg in 150 to 350 mm fish) and protection of human health (0.3 mg/kg for typical size and species of fish consumed). Urban runoff sources were given mass targets of 11 kg/yr to be achieved within 10 years and 7.2 kg/yr to be achieved within 20 years. The loads represent reductions of 21% and 49% from the existing load of 14 kg/yr. The final load reduction for urban runoff in the Guadalupe River watershed was taken from the San Francisco Bay mercury TMDL.

Although no load reductions were assigned to WWTPs in this TMDL (no point sources were identified as contributors of mercury loadings to the Guadalupe River), WWTPs discharging to San Francisco Bay are subject to the cap on their current load.

Las Tablas Creek/Lake Nacimiento (CRWQCB, 2002) – Like the other TMDLs reviewed for this project, the Las Tablas Creek/Lake Nacimiento TMDL had no point sources. It is of interest, however, in that the sediment target was set using the NOAA-PEL (Probably Effects Level) of 0.486 mg/kg. This sediment quality guideline was established by NOAA to help screen for *toxicity* effects for sediment. It therefore has little or no relevance to bioaccumulation or to fish tissue levels and ultimately to the protection of wildlife or human health. The report states that while it is not known whether achieving the 0.486 mg/kg PEF will result in the achievement of the water column target of 50 ng/L, “this value seems to be a reasonable point from which to start”.

Sulphur Creek, CA (CEPA, 2007) – The Sulphur Creek TMDL is focused completely on natural and mining sources. The allocation for reduction is based on the Cache Creek TMDL (Sulphur Creek is a tributary) and ultimately aimed at reducing MeHg concentrations in fish tissue.

Walker Creek, CA (CRWQCB, 2006a) – Walker Creek is solely impacted by natural and mining sources of mercury. Fish tissue targets are set at levels identical to those previously identified for the Guadalupe River for protection of wildlife. Sediment goals are set at 0.5 mg/kg mercury in sediment and the California Toxics Rule criterion of 50 ng/L in the water column is adopted for the protection of human health. An MeHg criterion of 0.04 ng/L for the protection of wildlife is also derived by dividing the 0.05 mg/kg target in fish tissue by a "nationwide" BAF value published by EPA of 1,300,000. A suspended sediment target of 0.5 mg/kg was set to insure the attainment of the 50 ng/L in water for protection of human health and a one-hour acute toxicity level of 2.4 ug/L for protection of aquatic organisms.

B.4.5 U.S. EPA Region 10

Stillaguamish River, WA (WDOE, 2005) – The mercury portion of this TMDL was focused on nonpoint sources in the watershed, as sources are all thought to be natural. There were three measurements of T-Hg from the Arlington, WA WWTP, ranging from 3.4 to 9.6 ng/L. It was observed on the one day when there were concurrent measurements of mercury in the WWTP effluent and in the receiving water, the T-Hg in the effluent was lower than in the receiving water. From a correlation analysis of T-Hg and TSS in samples from the river, it was concluded that TSS would need to be limited to 65 mg/L in order to limit T-Hg to the water quality criterion of 12 ng/L. The TMDL called for the implementation of erosion controls to limit TSS and therefore in-stream T-Hg concentrations.

B.5 The Importance of Mercury TMDLs to the Practice of Wastewater Treatment

Mercury TMDLs have been and will continue to exert and influence on wastewater treatment in the United States. This becomes obvious when comparing the typical WWTP effluent with the water quality targets established through the TMDL process. For instance, Chapter 3.0 provides a table showing levels of T-Hg discharges by wastewater treatment plants ranging from a few to several hundred ng/L. Appendix A presents a figure showing a range of total mercury in effluents from advanced wastewater treatment plants in CA ranging from a few to hundreds of ng/L. A recent study performed by NACWA, also discussed in Appendix A, showed T-Hg levels in WWTP effluents of up to 80 ng/L. Comparison of effluent discharge levels with the T-Hg water quality targets established by mercury TMDLs across the country (0.5 to 41 ng/L from Table 7-1 above) demonstrates that in a given waterbody, an individual WWTP may have serious difficulty being in compliance if water quality targets for T-Hg are adopted as end-of-pipe requirements. The same is true for MeHg where effluent concentrations may range as high as 2 to 3 ng/L and typical MeHg water quality targets from TMDLs are typically less than 0.5 ng/L (Table B-1). A complicating factor is that treatment of wastewater to "single-digit" ng/L levels or lower is expensive and in some cases may not technologically feasible. It is still not clear why some plants appear to achieve mercury removal to these levels with conventional technologies while other plants with similar treatment processes do not. Most emerging technologies for removal of mercury have not been tested in the high-volume low concentration environments of municipal WWTPs and are unproven in this context.

References

Abu-Saba, K.E., Leng, J., Tellefson, W., McCall, J., O'Brien, A., Pirodini, T., Paulucci, M., Fry, V., Gittings, S., Hartinger, C., Grovhoug, T., and Dunham, T. 2005. *A regional assessment of methylmercury discharges from municipal treatment plants in California's Central Valley.* Proceedings of Water Environment Federation 78th Annual Technical Exhibition (WEFTEC) October 29 – November 2, 2005, Washington, D.C.

Adams, N.W.H. and Kramer, J.R. 1999. *Environ. Toxicol. Chem.* **18**: 2667–2673.

ADA. 2007. American Dental Association, Best management practices for amalgam waste. www.ada.org/prof/resources/topics/amalgam_bmp.asp.

ADEQ and Tetra-Tech 2005. *Draft Final Mercury TMDL for Alamo Lake, AZ.* Arizona Department of Environmental Quality. December.

Allan, C.J. and Heyes, A. 1998. *Water Air Soil Poll.* **105**: 573-592.

Ansari, M.H., Deshkar, A.M., Kelkar, P.S., Dharmadhikari, D.M., Hasan, M.Z., and Paramasivam R. 1999. *Water Sci. Technol.* **7**: 109-116.

Asano, T. and Tchobanoglous, G. 1987. *Municipal wastewater treatment and effluent utilization for irrigation.* Paper prepared for the Land and Water Development Division, FAO**,** Rome.

Ashley, R.P., Rytuba, J.J., Rogers, R., Kotlyar, B.B., and Lawyer, D. 2002. *Preliminary report on mercury geochemistry of placer gold dredge tailings, sediments, bedrock, and waters in the Clear Creek restoration area, Shasta County, CA.* USGS. Open-file Report 02-401.

Babiarz, C.L., Hurley, J.P., Benoit, J.M., Shafer, M.M., Anders, A.W., and Webb, D.A. 1998. *Biogeochem.* **41**: 237-257.

Bae, W., Mehra, R.K., Mulchandani, A., and Chen, W. 2001. *Appl. Environ. Microbiol.* **67**: 5335-5338.

Balcom, P.H., Fitzgerald, W.F., Vandal, G.M., Lamborg, C.H., Rolfhus, K.R., Langer, C.S., and Hammerschmidt, C.R. 2004. *Mar. Chem.* **90:** 53-74.

Balogh, S.J., Huang, Y., Offerman, H.J., Meyer, M.L., and Johnson, D.K. 2003. *Sci. Total Environ.* **304**: 305-313.

Balogh, S.J., Meyer, M.L., and Johnson, D.K. 1998. *Environ. Sci. Technol.* **32**: 456-462.

Balogh, S.J., Nollet, Y.H., and Swain, E.B. 2004. *Environ. Sci. Technol.* **38**: 4921-4927.

Bloom, N.S. and Fitzgerald, WF. 1988. *Anal Chim Acta* **208**: 151-161.

Balogh, S. and Liang, L. 1995. *Water, Air, and Soil Pollution* **80**, 1181-1190.

Barkay, T., Gillman, M., and Turner, R.R. 1997. *Appl. Environ. Microbiol.* **63**: 4267-4271.

Barr Engineering Company. 2001. *Substance flow analysis of mercury in products.* Prepared for Minnesota Pollution Control Agency. Mineeapolis, MN. August 15.

Barron-Zambrano, J., Laborie, S., Viers, Ph., Rakib, M., and Durand, G. 2002. *Desalinisation.* **144**: 201-206.

Benoit, J., Gilmour, C., Heyes, A., Mason, R.P., and Miller, C. 2003. In: Chai Y, Braids OC, eds. *Biogeochemistry of environmentally important trace elements.* ACS, Washington, D.C., 262-297.

Benoit, J.M., Gilmour, C.C., Mason, R.P., Reidel, G.S., and Reidel, G.F. 1998. *Biogeochem.* **40**:249-265.

Benoit, J.M., Mason, R.P., Gilmour, C.C., and Aiken, G.R. 2001. *Geochim Cosmochim Acta* **65**: 4445-4451.

Benoit, J.M., Gilmour, C.C., Mason, R.P., Heyes A. 1999a. *Environ Sci Technol* **33**: 951-957.

Benoit, J.M., Mason, R.P., and Gilmour C.C. 1999b. *Enviro. Toxicol Chem* **18**: 2138-2141.

Bernhardsson, M. 2006. *Methylmercury in runoff from forested catchments: characterisation of three catchments prior to logging.* Department of Environmental Assessment Swedish University of Agricultural Sciences, Uppsala. Rapport 2006:2.

Bhandare, A.A. and Argekar, A.P. 2002. *J. Chem. Technol. Biotechnol.* **77**: 811-816.

Bishop, K., Lee, Y-H, Pattersson, C., and Allard, B. 1995. *Water Air Soil Poll.* **80**: 445-454.

Bloom, N.S. and Watras, C.J. 1989. *Sci. Total Environ.* **87/88**: 199-207.

Bloom, N.S., Gill, G.A., Cappellino, S., Dobbs, C., McShea, L., Driscoll, C., Mason, R., and Rudd, J.W.M. 1999. Environ. Sci. Technol. **33**:7-13.

Bodaly, R.A., Rudd, J.W.M., and Flett, R.J. 1998. *Biogeochem.* **40**: 279-291.

Bodaly, R.A., Rudd, J.W.M., Fudge, R.J.P., and Kelly, C.A. 1993. *Can. J. Fish Aquat. Sci.* **50**: 980-987.

Bonzongo, J.-C.J., Heim, K.J., Chen, Y., Lyons, W.B, Warwick, J.J., Miller, G.C., and Lechler, P.J. 1996. *Environ. Toxicol. Chem.* **15**: 677-683.

Bonzongo, J.-C.J. and Lyons, W.B. 2004. *Ambio.* **33**: 328-333.

Bowles, K.C., Bell, R., Ernste, M.J., Kramer, J.R., Manolopoulos, H., and Ogden, N. 2002. *Environ. Toxicol. Chem.* **21** : 693-699.

Boyd, C.E. 1979. *Water quality in warm water fish ponds.* Auburn Univ., Agriculture Experiment Station, AL. 359 pp.

Brigham, M.E., Krabbenhoft, D.P., Olson, M.L., and DeWild, J.F. 2002. *Water Air Soil Poll.* **138**: 61-78.

Callahan, J.E., Miller, J.W., and Craig, J.R. 1994. *Appl. Geochem.* **9**: 235-241.

Campbell, K.R., Ford, C.J., and Levine, D.A. 1998. *Environ. Toxicol. Chem.* **17**: 1191-1198.

Canario, J., Vale, C., and Caetano, M. 2005. *Marine Poll. Bull.* **50**: 1121-1145.

Canham, C.D., Pace, M.L., Papaik, M.J., Primack, A.G.M., Roy, K.M., Maranger, R.J., Curran, R.P., and Spada, D.M. 2004. *Ecol. Applic.* **14**: 839-854.

Cossa, D., Coquery, M., Gobeil, C., Martin, J.-M.. 1996. Mercury fluxes at the ocean margins. In *Global and Regional Mercury Cyles*, Baeyens et al. (Eds), NATO ASI Series, Kluwer Academic Press, pp 229-247.

CEP. 2003. *Mercury Management by Bay Area Wastewater Treatment Plants.* Clean Estuary Partnership. Task 4.5 Report.

CEPA 2005. *Sacramento – San Joaquin Delta Estuary TMDL for total and methyl mercury.* Staff report. Draft report. California Environmental Protection Agency, Regional Water Quality Control Board, Central Valley Region. August.

CEPA 2007. *Sulphur Creek TMDL for mercury.* Final staff report. California Environmental Protection Agency, Regional Water Quality Control Board, Central Valley Region, Rancho Cordova, CA. January.

Chen, C. and Folt, C.L. 2005. *Environ. Sci. Technol.* **39**: 115-121.

Chen, D. and Ray, A.K. 2001. *Chem. Engr. Sci.* **56**: 1561-1570.

Chen, Y., Bonzongo, J.C., and Miller, G.C. 1996. *Environ. Poll.* **92**: 281-287.

Chen, Z. and Mayer, L.M. *Mar. Ecol. Prog. Ser.* **176**: 139-151, 1999b.

Cogbill, C.V. and Likens, G.E. 1974. *Water Resour. Res.* **10**: 1133.

Colman, J.A., Waldron, M.C., Breault, R.F., and Lent, R.M. 1999. *Distribution and transport of total mercury and methylmercury in mercury-contaminated sediments in reservoirs and wetlands of the Sudbury River, east-central Massachusetts.* USGS. Water-Resources Investigations Report 99-4060.

Compeau, G. and Bartha, R. 1984. *Appl. Environ. Microbiol.* **48**: 1203-1207.

Conaway, C.H., Squire, S., Mason, R.P., and Flegal, A.R. 2003. *Marine Chem.* **80**: 199-225.

CRWQCB 2002. *Las Tablas Creek and Lake Nacimiento Mercury Total Maximum Daily Load for Mercury.* Draft. California Regional Water Quality Control Board, Central Coast Region, San Luis Obispo, CA. September 25.

CRWQCB 2006. *Guadalupe River watershed mercury total maximum daily load (TMDL) project report.* California Regional Water Quality Control Board, San Francisco Bay Region. January.

CRWQCB 2006a. *Total maximum daily load for mercury in the Walker Creek watershed.* Staff report. California Regional Water Quality Control Board, San Francisco Bay Region. November 20.

Cyr, P.J, Suri, R.P.S., and Helmig, E.D. 2002. *Water Research.* **36**: 4725-4734.

Deng, X. and Wilson, D.B. 2003. *Appl. Microbiol. Biotechnol.* **56**: 276-279.

Di Pasquale, M., Stewart, R., Fisher, N., Pickhardt, P., Mason, R.P., Heyes, A., and Windham-Meyes, L. 2007. *Evaluation of Mercury Transformations and Trophic Transfer in the San Francisco Bay/Delta: Identifying Critical Processes for the Ecosystem Restoration Program.* Draft Final Report for Project #ERP-02-P40, California Bay-Delta Authority, Sacramento, CA.

Domagalski, J. 2001. *Appl. Geochem.* **16**: 1677-1691.

Domagalski, J.L., Alpers, C.N., Slotten, D.G., Suchanek, T.H., and Ayers, S.M. 2004. *Sci. Tot. Environ.* **327**: 215-237.

Downs, S.G., MacLeod, C.L., and Lester, J.N. 1998. *Water Air Soil Poll.* **108**: 149-187.

Drott, A., Lambertsson, L., Beorn, E., and Skyllberg, U. 2007. *Environ. Sci. Technol.* **41**: 2270-2276.

Dujardin, M.C., Caze, C., and Vroman, I. 2000. *Reactive and Functional Polymers.* **43**: 123-132.

Eckenfelder, W.W. 1970. *Water quality engineering for practicing engineers.* Barnes & Noble, Inc. New York.

Effler, S.W. and Bloom, N.S. 1990. *Water Air Soil Poll.* **53**: 251-265.

Environmental Services Department (ESD) 2007. *San Jose/Santa Clara Water Pollution Control Plant mercury fate and transport study.* Report.

EPRI 2003. *Implementation of the United States Environmental Protection Agency's methylmercury criterion for fish tissue.* Electric Power Research Institute, Palo Alto, CA Technical Report #1005520.

EPRI. 2003. *Factors affecting the predicted response of fish mercury concentrations to changes in mercury loading.* Electric Power Research Institute, Palo Alto, CA. Technical Report #1005521.

EPRI 2006. *Mercury TMDLs – significance to the power industry and guidance for individual power plants.* Electric Power Research Institute, Palo Alto, CA. Technical Report #1010102.

Feng, X., Liu, J., Fryxell, G.E., and others. A997. *Self-Assembled Mercaptan on Mesoporous Silica (SAMMS) technology of mercury removal and stabilization.* Battelle Pacific Northwest Laboratory, Richland, WA. Technical report No. PNNL-11691.

Filali, B.K., Taoufik, J., Zeroual, Y., Dzairi, F.Z., Talbi, M., and Blaghen, M. 2000. *Current Microbiol.* **41**: 151-156.

Fitzgerald, W.F. and Mason, R.P. 1997. *In:* A. Sigel and H. Sigel [eds,]. *Mercury and Its Effects on Environment and Biology.* Marcel Dekker, NY. pp. 53-111.

Fitzgerald, W.F., Mason, R.P., and Vandal, G.M. 1991. *Water Air Soil Poll* **56**: 745-767.

Foe, C.G. 2003. *Mercury Mass Balance for the Freshwater Sacramento-San Joaquin Bay-Delta Estuary.* Final report submitted to the CALFED Bay-Delta Program for the project: An Assessment of the Ecological and Human Health Impacts of Mercury in the Bay-Delta Watershed (Task 1A). California Regional Water Quality Board, Central Valley Region, Sacramento, CA.

Foster, G.R. 1980. *Soil erosion modeling: special considerations for nonpoint pollution evaluation of field-sized areas. In: Environmental Impact of Nonpoint Source Pollution.* Overcash, M.R. and Davidson, J.M., Eds. Ann Arbor Science Publishers, Inc. Ann Arbor, MI.

Fryxell, G.E., Mattigod, S.V., Lin, Y., Wu, H., Fiskum, S., Parker, K., Xeng, F., Yantasee, W., Zemanian, T.S., Addleman, R.S., Liu, J., Kemner, K., Kelly, S., and Feng, X. 2007. *J. Mater. Chem.* **17**: 2863-2874.

FTN Associates. 2002. *TMDLs for segments listed for mercury in fish tissue for selected Arkansas watersheds.* Prepared for US Environmental Protection Agency, Region VI, Waste Management Section, Dallas, TX. September 17.

FTN Associates 2007. *Mercury TMDL for Spring Lake In Yell County Arkansas.* Final. Prepared for EPA Region 6, Water Quality Protection Division, Permits, Oversight, and TMDL Team, Dallas, TX. Contract No. 68-C-02-108, Task Order #127. January 16.

Gagnon, C., Pelletier, E., Mucci, A., and Fitzgerald, W.F. 1996. *Limnol. Oceanogr.* **41**: 428-434.

Galloway, J.N., Likens, G.E., Keene, W.C., and Miller, J.M. 1982. *J. Geophys. Res.* **87**: 8771-8786.

Ganguli, P.M., Mason, R.P., Abu-Saba, K.E., Anderson, R.S., and Flegal, A.R. 2000. *Environ. Sci. Technol.* **34**: 4773-4779.

Garcia, E., Amyot, M., and Ariya, P.A. 2005.. *Geochim Cosmochim Acta* **69**: 1917-1924.

Gash, A.E., Spain, A.L., Dysaleski, L.M., Flaschenreim, C.J., Kalaveshi, A., Dorhout, P.K., and Strauss, S.H. 1998. *Environ. Sci. Technol.* **32**: 1007-1012.

Gergel, S.E., Turner, M.G., and Krantz, T.K. 1999. *Ecol. Applic.* **9**: 1377-1390.

Giddings, E.M., Hornberger, M.I., and Hadley, H.K. 2001. Trace-metal concentrations in sediment and water and health of aquatic macroinvertebrate communities of streams near Park City, Summit County, Utah. USGS. Water-Resources Investigations Report 01-4213.

Gill, G.A., Bloom, N.S., Cappallino, S., Driscol,l C.T., Dobbs, C., McShea, L., Mason, R., and Rudd, J.W.P. 1999. *Environ. Sci. Technol.* **33**: 633-669.

Gill, G.A., Choe, K.Y., Lehman, R., and Han, S. 2003. Sediment-Water Exchange and Estuarine Mixing Fluxes of Mercury and Monomethylmercury in the San Francisco Bay Estuary and Delta. Final report submitted to the CALFED Bay-Delta Program for the project: An Assessment of the Ecological and Human Health Impacts of Mercury in the Bay-Delta Watershed (Task 4B). Laboratory for Oceanographic and Environmental Research, Texas A&M University, Galveston, TX.

Gilmour, C.C. and Bloom, N.S. 1995. *Water Air Soil Poll.* **80**: 799-803.

Gilmour, C.C. and Henry, E.A. 1991. *Environ. Poll.* **71**: 131-169.

Gilmour, C.C. and Riedel, G.S. 2000. *Arch. Environ. Contam. Toxicol.* **39** : 53-59.

Glass, G.E. and Sorensen, J.A. 1999. *Environ. Sci. Technol.* **33**: 3303-3312.

Golding, G.R., Kelly, C.A., Sparling, R., Loewen, P.C., Rudd, J.W.M., and Barkay, T. *Limnol. Oceanogr.* 47: 967-975.

Gordon, J.D., Schild D.E., Capesius J.P., and Slaughter C,B. 2005. *Surface water-quality and water-quantity data from selected urban runoff-monitoring sites at the Rocky Mountain Arsenal,*

Commerce City, Colorado, water years 1988-2004. USGS Scientific Investigation Report 2005-5214.

Granato, G.E. and Smith, K.P. 1999. *Estimating concentrations of road-salt constituents in highway-runoff from measurements of specific conductance.* USGS. Water Resources Investigation Report 99-4077.

Gray, J.E., Theodorakos, P.M., Bailey, E.A., and Turner, R.R. 2000. *Sci. Total Environ.* **260**: 21-33.

Gray, J.E. 2003. *Leaching, transport, and methylation of mercury in and around abandoned mercury mines in the Humboldt River basin and surrounding areas, Nevada.* USGS. U.S. Geological Survey Bulletin 2210-C.

Gray, J.E., Hines, M.E., Higueras, P.L., Adatto, I., and Lasorsa, B.K. 2004. *Environ. Sci. Technol.* **38**: 4285-4292.

Grigal, D.F. 2002. *Environ. Rev.* **10**: 1-39.

Gromaire-Mertz, M.C., Garnaud, S., Gonzalez, A., and Chebbo, G. 1999. *Wat. Sci. Tech.* **39**: 1-8.

Gunther, A.J., Salop, P., Bell, D., Feng, A., Wiegel, J., and Wood, R. 2001. Initial characterization of PCB, mercury, and PAH contamination in the drainages of western Alameda County, CA. Alameda Clean Water Program. March.

Hall, B.D., Manolopoulos, H., Hurley, J.P., Schafer, J.J., St. Louis, V.L., Kenski, D., Graydon, J., Babiarz, C.L., Cleckner, L.B., and Keeler, G.J. 2005. *Atmos. Environ.* **39**: 7557-7569.

Hammerschmidt, C.R. and Fitzgerald, W.F. 2004. *Environ. Sci. Technol.* **38**: 1487-1495.

Hammerschmidt, C.R., Lamborg, C.H., and Fitzgerald, W.F. 2007. *Atmos. Environ.* **41**: 1663-1668.

Harris, R.C., Rudd, J.W.M., Amyot, M., Babiarz, C., Beaty, K.G., Blanchfield, P.J., Bodaly, R.A., Branfireun, B.A., Gilmour, C.C., Graydon, J.A., Heyes, A., Hintelmann, H., Hurley, J.P., Kelly, C.A., Krabbenhoft, D.P., Lindberg, S.E., Mason, R.P., Paterson, M.J., Podemski, C.L., Robinson, A., Sandilands, K.A., Southworth ,G.R., St. Louis, V.L., and Tate, M.T. 2007. Whole-ecosystem study shows rapid fish-mercury response to changes in mercury deposition. Proceedings of the National Academy of Sciences of the United States of America 104(42): 16586-16591.

Heim, W.A., Coale, K., and Stephenson, M. Undated. *Methyl and total mercury spatial and temporal trends in surficial sediments of the San Francisco Bay-Delta. Assessment of ecological and human health impacts of mercury in the Bay-Delta watershed.* CALFED Bay-Delta Mercury Project Final Report.

Heim, W.A., Coale, K.H., Stephenson, M., Choe, K.-.Y., Gill, G.A., and Foe, C. 2007. *Environ. Sci. Technol.* **41**: 3501-3507.

Hellyer, G. 2000. *Ten Mile River Watershed-Massachusetts. An assessment of sediment chemistry and ecotoxicity.* United States Environmental Protection Agency, New England Regional Laboratory. North Chelmsford, MA. October 5.

Hendry, C.D. and Brezonik, P.L. 1980. *Environ. Sci. Technol.* **14**: 843-849.

Herrin, R.T., Andren, A.W., Shafer, M.M., Armstrong, D.E. 2001. *Environ. Sci. Technol.* **35** : 1959-1966.

Heyes, A., Mason, R.P., Kim, E.-H., and Sunderland, E. 2006. *Mar. Chem.* **102**: 134-147.

Heyes, A., Miller, C., and Mason, R.P. 2004. *Mar. Chem.* **90**: 75-89.

Hines, M.E., Bonzongo, J.J., Barkay, T., Horvat, M. and Faganelli, J. 2001. *Mercury contamination and biogeochemical cycling associated with the historic Idrija mining area of Slovenia.* EOS Trans. AGU, 82(47). Fall Meet. Suppl., Abstract H52F-11.

Hintelmann, H. and Harris, R. 2004. *Marine Chem.* **90**: 165-174.

Hollerman, W., Holland, L., Ila, D., Hensley, J., Southworth, G., Klasson, T., Taylor, P., Johnston, J., and Turner, R. 1999. J. Hazard. Mater. **10**: 193-203.

Holz, J., Kreutzmann, J., Wilken, R.-D., and Falter, R. 1999. *Appl. Organomettalic Chem.* **13**: 789-794.

Horvat, M., Nolde, N., Fajon, V., Vesna, J., Logar, M., Logen, S., Jacimovic, R., Falnoga, I., Liya, Q., Faganelli, J., and Drobne, D. 2003. *Sci Total Environ.* **304**: 231-256.

Hsu, H. 2004. *Mercury biogeochemistry and removal during municipal wastewater treatment.* Technical report for the City of San Jose, Environmental Services Department. August.

Hsu, H. and Sedlak, D.L. 2003. *Environ Sci Technol* **37**: 2743-2749.

Hsu-Kim, H. and Sedlak ,D.L. 2005. *Environ Sci Technol* **39**: 4036-4041.

Hudson, J.J., Dillon, P.J., and Somers, K.M. 2003. *Hydrol. Earth Sys. Sci.* **7**: 390-398.

IDEQ 2004. *Draft implementation guidance for the Idaho mercury water quality criterion.* Idaho Department of Environmental Quality. Boise, ID. July, 2004.

Jay, J.A., Murray, K.J., and Gilmour, C.C., et al., 2002. *Appl Environ Microbiol.* **68**: 5741-5745.

Jay, J.A., Morel, F.M.M., and Hemond, H.F. 2000. *Environ. Sci. Technol.* **34**: 2196-2200.

Jeremiason, J.D., Engstrom, D.R., Swain, E.B., Nater, E.A., Johnson, B.M., Almendinger, J.E., Monson, B.A., and Kolka, R.K. 2006. *Environ. Sci. Technol.* **40**: 3800-3806.

Jones, A.B., Slotten, D.G., Foe ,C., and Domagalski, J. 1996. *Mercury effects, sources and control measures.* San Francisco Estuary Regional Monitoring Program. RMP Contribution #20. September.

Jonsson, H.H. and Vonnegut, B. 1991. *J. Appl. Meteor.* **30**: 1220-1227.

Junge, C.E. and Werby, R.T. 1958. *J. Meteor.* **15**: 417-425.

Kamman, N.C., Chalmers, A., Clair, T.A., Major, A., Moore, R.B., Norton, S.A., and Shanley, J.B. 2005. *Ecotoxicol.* **14**: 101-111.

Kannan, K., Smith, R.G., Lee, R.F., Windom, H.L., Heitmuller, P.T., Macauley, J.M. and Summers, J.K. 2004. *Arch. Environ. Contam. Toxicol.* **34**: 109-118.

Kannan, N. and Rajakumar, A. 2001. *Toxicol. and Environ. Chem.* **84**: 7-19.

Karlsson, T. and Skyllberg, U. 2003. *Environ. Sci. Technol.* **37**: 4912-4918.

Kelly, C.A., Rudd, J.W.M., and Holoka, M.H. 2003. *Environ. Sci. Technol.* **37**: 2941-2946.

Kerin, E.J., Gilmour, C.C., Roden, E., Suzuki, M.T., Coates, J.D., and Mason, R.P. 2006. *Appl. Environ. Microbiol.* **72** : 7919-7921.

Kim, E.-H., Mason, R.P., Bergeron, C.M. (submitted). *Modeling methylmercury bioaccumulation in an estuarine environment: An examination of the major controlling factors.* Ecological Applications.

Kim, E.-H., Mason, R.P., Porter, E.T., Soulen, H.L. 2006. *Mar Chem* **102**: 300-315.

King, J.K., Gladden, J.B., Harmon, S.M. and Fu, T.T. 2001. *Mercury removal, methylmercury formation, and sulfate-reducing bacteria profiles in wetland mesocosms containing gypsum-amended sediments and Scirpus californicus.* Department of Energy Technical Report No. WSRC-TR-2001-00063.

Kinnetic Laboratories, Inc. 2002. Joint stormwater agency project to study urban sources of mercury, PCBs and organochlorine pesticides. April.

Kolka, R.K., Grigal, D.F., Nater, E.A., and Verry, E.S. 2001. *Soil Sci. Soc. Am. J.* **65**: 897-905.

Lambertsson, L. and Nilsson, M. 2006. *Environ. Sci. Technol.* **40**: 1822-1829.

Lamborg, C.H., Fitzgerald, W.F., Vandal, G.M., and Rolfhus, K.R. 1995. *Water Air Soil Poll.* **80**: 189-198.

Lamborg, C.H., Tseng, C.-M., Fitzgerald, W.F., Balcom, P.H., and Hammerschmidt, C.R. 2003. *Environ Sci Technol* **37**: 3316-3322.

Lamborg, C.H., Fitzgerald, W.F., Skoog, A., and Visscher, P.T. 2004. *Mar. Chem.* **90** : 151-163.

Lanning, J.M. 1993. *Aqueous mercury treatment technology review for NPDES Outfall 49 Y-12 Plant.* U.S. Department of Energy, Oak Ridge TN. Report No. Y/DZ-1043.

Laurier, F. and Mason, R.P. 2007. *Mercury concentration and speciation in the coastal and open ocean boundary layer*. JGR-Atmos. 112, D06302, doi: 10.1029/2006JD007320.

Lawrence, A.L, McAloon, K.M., Mason, R.P., and Mayer, L.M. *Environ. Sci. Technol.* **33**: 1871-1876, 1999.

Lawson, N.M. and Mason, R.P. 1998. *Biogeochem* **40**: 235-247.

Lawson, N.M., Mason, R.P., and Laporte, J.-M. 2001 Wat. Res., **35**:501-515.

Lee, J.-S., Lee, B.-G., Luoma, S.N., Choi, H.J., and Koh, C.-H. 2000. *Environ. Sci. Technol.* **34**: 4517-4523.

Lee, Y.-A. and Iverfelt, A. 1991. *Water Air Soil Poll.* **56**: 309-321.

Leermakers, M., Galletti, S., De Galan, S., Brion, N., and Baeyens, W. 2001. *Marine Chem.* **75**: 229-248.

Lehmann, C.M.B., Bowersox, V.C., Larson, R.S., and Larson, S.M. 2007. *Water Air Soil Pollut: Focus* **7**: 59-66.

Leigh, D.S. 1997. *Environ. Geol.* **30**: 244-251.

Lillie, R.A and Mason, J.W. 1983. Limnological characteristics of Wisconsin lakes. Tech. Bull. Wisconsin Department of Natural Resources, No. 138, Madison, WI. Hine L, Ed. 116 p.

Lindsay, W.L. 1979. Chemical equilibria in soils. John Wiley & Sons, Inc., New York.

Loredo, J., Ordonez, A., Charlesworth, S., and De Miguel, E. 2003. Environ. Geochem. Health. **25**: 307-323.

Luther, G.W., Theberge S.M., Rickard D.T. 1999. *Geochim. Cosmochim. Acta* **63** : 3159-3169.

Lynch, J.A., Bowersox, V.C., and Grimm, J.W. 1996. *Trends in Precipitation Chemistry in the United States, 1983-94: An Analysis of the Effects in 1995 of Phase I of the Clean Air Act Amendments of 1990, Title IV.* USGS. Open-File Report 96-0346.

Macalady, J.L., Mack, E.E., Nelson, D.C., and Scow, KM. 2000. *Appl. Environ. Microbiol.* **66**: 1479-1488.

Maglambayan, B.V., Murao, S., Corpus, T.J.C., Sera, K., Futatsugawa, S., and Tsuji, M. 2005. *Mercury contamination associated with small-scale gold mining in the upper Ambalanga River, Benguet, Philippines, from river sediment sampling.* Asia-Pacific Learning Event, Philippines. June 7-12

Maret, T.R. and Skinner, K.D. 2000. *Concentrations of selected trace elements in fish tissue and streambed sediment in the Clark Fork-Pend Oreille and Spokane River basins, Washington, Idaho and Montana, 1988.* USGS. Water-Resources Investigations Report 00-4159.

MASCO. 1997. *Report of the MASCO/MWRA Hospital Mercury Workgroup. Technology Identification Subgroup. Medical Academic and Scientific Community, Inc.,* Boston, MA.

Mashburn, S.L. and Sughru, M.P. 2004. *Chloride in ground water and surface waster in the vicinity of selected surface-water sampling sites of the beneficial use monitoring program of Oklahoma, 2003.* USGS Scientific Investigations Report 2004-5060.

Mason, R.P. 1991. *The chemistry of mercury in the equatorial Pacific Ocean.* PhD dissertation, University of Connecticut, Storrs, CT, pp. 305.

Mason, R.P. 2002. *The bioaccumulation of mercury, methylmercury and other toxic elements intopelagic and benthic organisms.* pp. 127-149. *In*: M.C. Newman, M.H. Roberts, and R.C. Hale [eds.], *Coastal and Estuarine Risk Assessment*, CRC/Lewis Publ.

Mason, R.P., Abbott, M.L., Bodaly, R.A., Bullock, O.R., Driscoll, C.T., Evers ,D., Lindberg, S.E., Murray, M., Swain, E.B. 2005. *Environ Sci Technol* **39**: 14A-22A.

Mason, R.P. and Benoit, J.M. 2003. *Organomercury compounds in the environment.* pp. 57-99. *In:* P. Craig [ed.], *Organometallics in the Environment*, John Wiley & Sons, New York.

Mason, R.P, Heyes, D., and Sveinsdottir, A. 2006. *Arch. Environ. Contam. Toxicol.* **51**: 425-437.

Mason, R.P. and Lawrence, A.L. 1999. *Environ. Toxicol. Chem.* **18**: 2438-2447.

Mason, R.P., Lawson, N.M. and Sullivan, K.A. 1997. *Atmos. Environ.* **31**: 3541-3550.

Mason, R.P., Lawson, N.M., Lawrence, A.L., Leaner, J.J., Lee, J.G., and Sheu, G.-R. 1999. *Mar. Chem.* **65**:77-96.

Mason, R.P., Lawson, N.M., and Sheu, G.R. 2000. *Atmos. Environ.* **34**: 1691-1701.

Mason, R.P., Reinfelder, J.R., and Morel, F.M.M.. 1996. *Environ Sci Technol* **30**:1835-1845.

Mason, R.P., Reinfelder, J.R., and Morel, F.M.M. 1995. *Water Air Soil Poll* **80**: 915-921.

Mason, R.P. and Sullivan, K.A. 1998. *Wat. Res.* **32**: 321-330.

Mason, R.P. and Sveinsdottir, A. 2003. *Mercury and methylmercury concentrations in water and largemouth bass in Maryland reservoirs.* Final Report. Maryland Department of Natural Resources. Annapolis, MD. October.

Mason, R.P., Fitzgerald, W.F., and Vandal, G.M. 1992. *J Atmos Chem* **14**: 489-500.

McAlear, J.A. 1996. *Concentrations and fluxes of total mercury and methylmercury within a wastewater treatment plant.* Master's Thesis. Enironmental Engineering, Syracuse University.

McAloon, K.M. and Mason, R.P. 2003. *Mar. Poll. Bull.* **46**: 1600-1608.

McGrath, J.A., Paquin, P.R., and DiToro, D.M. 2002. *Use of the SEM and AVS approach in predicting metal toxicity in sediments.* Fact Sheet on Environmental Risk Assessment, No 10. International Council on Mining and Metals, London, United Kingdom.

McKee, L., Leatherbarrow, J., and Eads, R. 2004. *Concentrations and loads of mercury, PCBs and OC pesticides associated with suspended sediments in the lower Guadalupe River, San Jose, California.* Contribution 86. San Francisco Estuary Institute. October.

McKee, L., Leatherbarrow, J., and Oram, J. 2005. *Concentrations and loads of mercury, PCBs and OC pesticides in the lower Guadalupe River, San Jose, California: water years 2003 and 2004.* Contribution 409. San Francisco Estuary Institute. July.

MDEQ 2004. *Mercury permitting strategy. Implementation of Method 1631 for fiscal years 2005 – 2009.* Michigan Department of Environmental Quality. May 14, 2004.

Mikac, N., Niessen, S., Ouddane, B., and Wartel, M. 1999. *Appl. Organomettalic Chem.* **13**: 715-725.

Miller, C.L., Mason, R.P., and Heyes, A. *Geochim Cosmochim Acta, submitted Sept 2006.*

Miller, C.L., Mason, R.P., Gilmour, C.C., and Heyes, A. 2007. *Environ. Toxicol. Chem.* **26**: 624-633.

Miller, C.L. 2005. *The role of organic matter in the dissolved phase speciation and solid phase portioning of mercury.* PhD Dissertation, University of Maryland, College Park, pp 159.

Mohan, D., Gupta, V.K., Srivastava, S.K., and Chander, S. 2000. *Colloids and Surfaces A: Physicochemical and Engineering Aspects.* **177**: 169-181.

Monteagudo, J.M. and Ortiz, M.J. 2000. *J. Chem. Technol. Biotechnol.* **75**: 767-772.

Moore, D.R.J., Pawlisz, A., Teed, R.S., Richardson, G.M., Allen, H.G., Thakali, S., Gibson, J.C., Hickey, B.C., Hill, J.R., Holmes, J., Ridal, J.J., Lean, D.S., Crow, J., Eskew, D., Holdsworth, G., and Little, J. 2003. *Developing ambient water quality criteria for mercury: A probabilistic site-specific approach.* Water Environment Research Foundation. 99-ECO-2.

Morel, F.M.M., Kraepiel, A.M.L., and Amyot, M. 1998. *Annu Rev Ecol Systemat* **29**:543-566.

Morel, F.M.M. and Hering, J.G. 1993. *Principles and applications of aquatic chemistry.* Wiley and Sons, NY.

MSDEQ 2004. *Pascagoula River phase one total maximum daily load for mercury.* Mississippi Department of Environmental Quality, TMDL/WLA Branch, Jackson, MS. October.

Munthe, J., Hultberg, H., and Iverfelt, A. 1995. *Water Air Soil Poll.* **80**: 363-371.

Najera, I., Lin, C.C., Kohbodi, G.A., and Jay, J.A. *Environ. Sci. Technol.* **39**: 3116-3120.

Namasivayam, C. and Kadirvelu, K. 1999. *Carbon.* **1**: 79-84.

Navarro, R.R., Sumi, K., Fujii, N., and Matsumura, M. 1996. *Water Research.* **30**: 2488-2494.

Nelson, E.A., Specht, W.L., and Knox, A.S. 2004. *Metal removal from process and stormwater discharges by constructed treatment wetlands.* Department of Energy Report No. WSRC-MS-2004-00763.

NEORSD. 2006. 2005 Southerly Wastewater Treatment Center Mercury Mass Balance Study. Northeast Ohio Regional Sewer District. November.

NEORSD. 2006a. Cuyahoga River Mercury Sampling Report. Northeast Ohio Regional Sewer District.

Novotny, V. and Chesters, G. 1981. *Handbook of nonpoint source pollution. Sources and management.* Van Nostrand Reinhold Company, New York.

NRC. 1993. *Managing Wastewater in Coastal Urban Areas.* National Research Council. National Academy Press. Washington, D.C.

ODEQ. 2004. *Willamette Basin TMDL. Chapter 3: Mercury TMDL.* Oregon Department of Environmental Quality. Salem, OR. September.

Ohio EPA 2000. *Permit guidance 10.* Final. Mercury variance guidance. Ohio EPA, Division of Surface Water Revision 0, June 23, 2000.

Okino, S., Iwasaki, K., Yagi, O., and Tanaka, H. *Biotechnology Letters.* **22**: 783-788.

Olinger, K.D. Undated. *Sediment quality assessment for the tidal Christina River basin. Vol. 1. Report of Findings.* Delaware Department of Natural Resources and Environmental Control. New Castle, DE.

Ordonez, A., Loredo, J., De Miguel, E., and Charlesworth, S. 2003. *Arch. Environ. Contam. Toxicol*. **44**: 160-170.

Pak, K.-R. and Bartha, R. 1998. *Appl. Environ. Microbiol.* **64**: 1013-1017.

Patton, G.W. and Crecelius, E.A. 2001. *Simultaneously extracted metals/acid-volatile sulfide and total metals in surface sediment from the Hanford reach of the Columbia River and the Lower Snake River.* Battelle Pacific Northwest National Laboratory, PNNL Report No. 13417.

Pavlogeorgatos, G.D., Thomaidis, N.S., Nikolaou, A.D., and Lekkas, T.D. 2006. Global NEST Jour. **8**: 61-67.

PerrinNadif, R., Dusch, M., and Koch, C., et al., 1996. *J Toxicol Environ Health* **48**: 107-119.

Pettersson, C., Bishop, K., Lee, Y.-H., and Allard, B. 1995. *Water Air Soil Poll.* **80**: 971-979.

Piani, R., Covelli, S., and Biester, H. 2005. *Appl. Geochem.* **20**: 1546-1559.

Pitt, R., Maestre A., and Morquecho R. 2004. *The National Stormwater Quality Database (NSQD, version 1.1).*

Polkowska, Z., Grynkiewicz, M., Zabiegala, B., and Namiesnik, J. 2001. *Polish J Environ Studies.* **10**: 351-363.

Porvari, P. and Verta, M. 2003. *Environ. Poll.* **123**: 181-191.

Poulain, A.J., Amyot, M., and Findlay, D., 2004. *Limnol Oceanogr* **49**: 2265-2275.

Pulido-Novocio, L., Kurimoto, Y., Aoyama, M., Seki, K., Doi, S., Hata, T., Ishihara, S., and Imimura, Y. 2001. *J. Wood Sci.* **47**: 159-162.

Ravichandran, M. 2004. *Chemosphere* **55**: 319-331.

Rearick, M.S. 2004. In situ measurement of sulphide in natural waters. MS Thesis, University of Maryland, College Park, pp. 83.

Rice, K.C. 1999. *Environ. Sci. Technol.* **33**: 2499-2504.

Rolfhus, K.R., Sakamoto, H.E., Cleckner, L.B., Stoor, R.W., Babiarz, C.L., Back, R.C., Monolopoulos, H., and Hurley, J.P. 2003. *Environ. Sci. Technol.* **37**: 865-872.

Rozan, T.F. and Luther, G.W. 2002. In *Environmental electrochemistry: analysis of trace element biogeochemistry,* Taillefert et al. (eds), ACS Symposium Series, Washington, DC, pp. 371-387.

Rozan, T.F., Lassman, M.E., Ridge, D.P., and Luther, G.W. 2000. *Nature* **406**: 879-882.

Rudd, J.W.M., Kelly, C.A., St. Louis, V., Hesslein, R.H., Furutani, A., and Holoka, M.H. 1986. *Limonol. Oceanogr.* **31**: 1267-1280.

RWQCB. 2004. *Mercury in San Francisco Bay. Total maximum daily load (TMDL) proposed basin plan amendment and staff report. Appendix C. Staff report.* California Regional Water Quality Control Board, San Francisco Bay Region. September 2, 2004.

Rytuba, J.J. 2000. *Sci. Total Environ.* **260**: 57-71.

Rytuba, J.J. 2002. Mercury geoenvironmental models. *Chapter J.* In: Progress on Geoenvironmental Models for Selected Mineral Deposit Types, *Robert R. Seal, RR II and Foley NK, Eds. U.S. Geological Survey Open-File Report 02-19.*

Schwarzenbach, R.P., Gschwend, P.M., Imboden, D.M. 1993. *Environmental organic chemistry*, Wiley and Sons, NY.

Sellers, P., Kelly, C.A., and Rudd, J.W.M. 1995. *Water Air Soil Pollut* **80**: 697-713.

Sequeira, R. and Lung, F. 1995. *Atmos. Environ.* **29**: 2439-2447.

Shafer, M.M., Overdier, J.T., and Armstrong, D.E. 1998. *Environ. Toxicol. Chem.* **17**: 630-641.

Shanley, J.B., Kamman, N.C., Clair, T.A., and Chalmers, A. 2005. *Ecotoxicol.* **14**: 125-134.

Shibi, I.G. and Anirudhan, T.S. 2002. *Ind. Eng. Chem. Res.* **41**: 5341-5352.

Skyllberg, U., Bloom, P.R., Qian, J., et al., 2006. *Environ Sci Technol* **40**: 4174-4180.

Sloan, J.J., Dowdy, R.H., Balogh, S.J., and Nater, E. 2001. *J. Environ. Qual.* **30**: 2173-2179.

Smith, D.S., Bell, R.A., Valliant, J., and Kramer, J.R. 2004. *Environ. Sci. Technol.* **38**: 2120-2125.

Soller, J., Stephenson, J., Olivieri, K., Downing, J., and Olivieri, A.W. 2005. *J. Environ. Mngmt.* **76**: 309-318.

Sreedhar, M.K. and Anirudhan, T.S. *J. Appl. Polymer Sci.* **75**: 1261-1269.

Stathopoulou, E., Dasenakis, M., and Scoulos, M. 2001. *Levels of mercury concentration in sediments of Saronikos Gulf.* Proceedings of the 7th International Conference of Environmental Science and Technology. Syros, Greece. September 3-6.

Steffan, R.J., Korthals, E.T. and Winfrey, M.R. 1988. *Appl. Environ. Microbiol.* **54**: 2003-2009.

St. Louis, V.L., Rudd, J.W.M., Kelly, C.A., Hall, B.D., Rolfhus, K.R., Scott, K.J., Lindberg, S.E., and Dong, W. 2001. *Environ. Sci. Technol.* **35**: 3089-3098.

Stumm, W. and Morgan, J.J. 1996. *Aquatic Chemistry*, 3rd Ed., John Wiley and Sons, NY

Suchanek, T.H., Slotton, D.G., Nelson, D.C., Ayers, S.M., Asher, C., Weynad, R., Liston, A., and Eagles-Smith, C. 2003. *Mercury Loading and Source Bioavailability from the Upper Cache Creek Mining Districts.* Final report submitted to the CALFED Bay-Delta Program for the project: An Assessment of the Ecological and Human Health Impacts of Mercury in the Bay-Delta Watershed (Task 5A). University of California, Davis, CA.

Takeuchi, F., Negishi, A., Maeda, T., Kamimura, K., and Sugio, T. 2003. *J. Biosci. Bioengr.* **95**: 239-244.

Taylor, P.A. and Klasson, K.T. 1999. Results of small-scale tests for removing mercury from ORNL process wastewater. Chemical Technology Division, Oak Ridge National Laboratory, U.S. Department of Energy. Report No. ORNL/TM-13763.

Thomas, M.A., Conaway, C.H., Steding, D.J., Marvin-DiPasquale, M., Abu-Saba, K.E., and Flegal, A.R. 2002. *Geochemistry: Exploration, Environment, Analysis.* **2**: 211-217.

Topping, B.R., Kuwubara, K.S., Marvin-DiPasquale, M.C., Agee, J.L., Kieu, L.H., Flanders, J.R., Parchaso, F., Hager, S.W., Lopez, C.B., and Krabbenhoft, D.P. 2004. *Sediment remobilization of mercury in South San Francisco Bay, California.* USGS. Scientific Investigations Report 2004-5196.

Tossell, J.A. 2001. *J. Phys. Chem. A* **105**: 935-941.

Tsai, P. and Hoenicke, R. 2001. *San Francisco Bay atmospheric deposition pilot project part 1: mercury.* San Francisco Estuary Institute. Contribution 72. July.

Ullrich, S.M., Tanton, T.W. and Abdrashitova, S.A. 2001. *Crit. Rev. Environ. Sci. Technol.* **31**: 241-293.

U.S. EPA. 1995. Final water quality guidance for the Great Lakes system: Final rule. *Fed. Regist.* **60**(56): 15366-15425 (March 23, 1995).

USDA. 1997. Water quality and agriculture. Status, conditions, and trends. Working Paper #16. Natural Resources Conservation Service. July.

U.S. EPA. 1996. United States Environmental Protection Agency, Office of Water (4305). EPA 823-B-96-007. June. 60p.

U.S. EPA. 1997. *Capsule report. Aqueous mercury treatment.* United States Environmental Protection Agency. Office of Research and Development. EPA/615/R-97/004. July.

U.S. EPA. 2000. Total maximum daily load (TMDL) development for total mercury and fish consumption guidelines in the middle and lower Savannah River watershed. US Environmental Protection Agency, Region IV, Atlanta, GA. December 8.

U.S. EPA. 2000a. Bayou Courtableau TMDL for sulfate. EPA Region 6. Dallas, TX.

U.S. EPA. 2001. Draft South Georgia mercury TMDL data report. United States Environmental Protection Agency, Science and Ecosystem Support Division, Ecological Assessment Branch, Athens, GA. July 20.

U.S. EPA. 2001a. Draft Middle Georgia mercury TMDL data report. United States Environmental Protection Agency, Science and Ecosystem Support Division, Ecological Assessment Branch, Athens, GA. August 20.

U.S. EPA. 2006. *Biosolids Technology Fact Sheet: Multistage Anaerobic Digestion*, Office of Water, Report # EPA 832-F-06-031.

Vaithiyanathan, P., Richardson, C.J., Kavanaugh, R.G., Craft, C.B., and Barkay, T. 1996. *Environ. Sci. Technol.* **30**: 2591-2597.

Van Veen, E., Burton, N., Comber, S., and Gardner, M. 2002. *Environ. Toxicol. Chem.* **21**: 275-280.

Von Canstein, H., Li, Y., Timmis, K.N., Deckwer, W.-D., and Wagner-Doebler, I. 1999. *Appl. Environ Mibrobiol.* **65**: 5279-5284.

Von Canstein, H., Li, Y., and Wagner-Doebler, I. 2001. *Biotechnol. Bioengr.* **74**: 212-219.

Von Canstein, H., Kelly, S., Li, Y., and Wagner Doebler, I. 2002. *Appl. Environ. Microbiol.* **68**: 2829-2837.

Wagner-Doebler, I., von Canstein, H., Timmis, K.N., and Deckwer, W.-D. 2000. *Environ Sci. Technol.* **34**: 4628-4634.

Wagner-Doebler, I., Luensdorf, H., Lubenhuesen, T., von Canstein, H.F., and Li, Y. 2000a. *Appl. Environ. Microbiol.* **66**: 4559-4563.

Wagner-Doebler, I. 2003. *Appl. Microbiol Biotechnol.* **62**: 124-133.

WDOE 2005. *Stillaguamish River watershed fecal coliform, dissolved oxygen, pH, mercury and arsenic total maximum daily load (water cleanup plan).* Submittal report. Publication no. 05-10-044. Washington State Department of Ecology, Water Quality Program, Northwest Regional Office, Bellevue and Environmental Assessment Program, Olympia, WA. April.

Whalin, L.M. and Mason, R.P. 2006. *Anal Chim Acta* **558**: 211-221.

Wiener, J.G., Krabbenhoft, D.P., Heinz, G.H., Scheuhammer, A.M. 2003. In: Hoffman, D.J., Rattner, B.A., Burton, G.A., Jr, Cairns, J. Jr., (eds), *Handbook of Ecotoxicology*, 2nd ed. CRC Press, Boca Raton, FL. p 409-463.

WHO. 1996. Chloride in drinking-water. In: Guidelines for drinking-water quality. 2nd ed. Vol.2. Health Criteria and Other Supporting Information. World Health Organization, Geneva.

WHO. 2004. Sulfate in Drinking-water. Background document for development of WHO guidelines for drinking-water quality. World Health Organization, Geneva, Switzerland.

Wood, M.L., Foe, C., and Cooke, J. 2006. *Sacramento – San Joaquin Delta Estuary TMDL for Methylmercury.* California Environmental Protection Agency, Regional Water Quality Control Board, Central Valley Region, Staff Report. June 2006, Draft Report for Scientific Peer Review.

WYA. 2007. *Advanced wastewater treatment capabilities for California Toxics Rule constituents and other trace toxics: evaluation of research needs.* West Yost Associates. Report of the Special Projects Group, Central Valley Clean Water Association. January.

Zeroual, Y., Moutaouakkil, A., and Blaghen, M. 2001. *Current Microbiol.* **43**: 322-327.

WASTEWATER UTILITY

Alabama
Montgomery Water Works & Sanitary Sewer Board

Alaska
Anchorage Water & Wastewater Utility

Arizona
Glendale, City of, Utilities Department
Mesa, City of
Peoria, City of
Phoenix Water Services Dept.
Pima County Wastewater Management
Safford, City of
Tempe, City of

Arkansas
Little Rock Wastewater Utility

California
Central Contra Costa Sanitary District
Corona, City of
Crestline Sanitation District
Delta Diablo Sanitation District
Dublin San Ramon Services District
East Bay Dischargers Authority
East Bay Municipal Utility District
El Dorado Irrigation District
Fairfield-Suisun Sewer District
Fresno Department of Public Utilities
Inland Empire Utilities Agency
Irvine Ranch Water District
Las Gallinas Valley Sanitary District
Las Virgenes Municipal Water District
Livermore, City of
Los Angeles, City of
Los Angeles County, Sanitation Districts of
Napa Sanitation District
Novato Sanitary District
Orange County Sanitation District
Palo Alto, City of
Riverside, City of
Sacramento Regional County Sanitation District
San Diego Metropolitan Wastewater Department, City of
San Francisco, City & County of
San Jose, City of
Santa Barbara, City of
Santa Cruz, City of
Santa Rosa, City of
South Bayside System Authority
South Coast Water District
South Orange County Wastewater Authority
Stege Sanitary District
Sunnyvale, City of
Union Sanitary District
West Valley Sanitation District

Colorado
Aurora, City of
Boulder, City of
Greeley, City of
Littleton/Englewood Water Pollution Control Plant
Metro Wastewater Reclamation District, Denver

Connecticut
Greater New Haven WPCA
Stamford, City of

District of Columbia
District of Columbia Water & Sewer Authority

Florida
Broward, County of
Fort Lauderdale, City of
Jacksonville Electric Authority (JEA)
Miami-Dade Water & Sewer Authority
Orange County Utilities Department
Pinellas, County of
Reedy Creek Improvement District
Seminole County Environmental Services
St. Petersburg, City of
Tallahassee, City of
Tampa, City of
Toho Water Authority
West Palm Beach, City of

Georgia
Atlanta Department of Watershed Management
Augusta, City of
Clayton County Water Authority
Cobb County Water System
Columbus Water Works
Fulton County
Gwinnett County Department of Public Utilities
Savannah, City of

Hawaii
Honolulu, City & County of

Idaho
Boise, City of

Illinois
Decatur, Sanitary District of
Greater Peoria Sanitary District
Kankakee River Metropolitan Agency
Metropolitan Water Reclamation District of Greater Chicago
Wheaton Sanitary District

Iowa
Ames, City of
Cedar Rapids Wastewater Facility
Des Moines, City of
Iowa City

Kansas
Johnson County Wastewater
Lenexa, City of
Unified Government of Wyandotte County/ Kansas City, City of

Kentucky
Louisville & Jefferson County Metropolitan Sewer District

Louisiana
Sewerage & Water Board of New Orleans

Maine
Bangor, City of
Portland Water District

Maryland
Anne Arundel County Bureau of Utility Operations
Howard County Bureau of Utilities
Washington Suburban Sanitary Commission

Massachusetts
Boston Water & Sewer Commission
Massachusetts Water Resources Authority (MWRA)
Upper Blackstone Water Pollution Abatement District

Michigan
Ann Arbor, City of
Detroit, City of
Holland Board of Public Works
Saginaw, City of
Wayne County Department of Environment
Wyoming, City of

Minnesota
Rochester, City of
Western Lake Superior Sanitary District

Missouri
Independence, City of
Kansas City Missouri Water Services Department
Little Blue Valley Sewer District
Metropolitan St. Louis Sewer District

Nebraska
Lincoln Public Works and Utilities Department

Nevada
Henderson, City of
Las Vegas, City of
Reno, City of

New Jersey
Bergen County Utilities Authority
Ocean County Utilities Authority
Passaic Valley Sewerage Commissioners

New York
New York City Department of Environmental Protection

North Carolina
Charlotte/Mecklenburg Utilities
Durham, City of
Metropolitan Sewerage District of Buncombe County
Orange Water & Sewer Authority
University of North Carolina, Chapel Hill

Ohio
Akron, City of
Butler County Department of Environmental Services
Columbus, City of
Metropolitan Sewer District of Greater Cincinnati
Montgomery, County of
Northeast Ohio Regional Sewer District
Summit, County of

Oklahoma
Oklahoma City Water & Wastewater Utility Department
Tulsa, City of

Oregon
Albany, City of
Clean Water Services
Eugene, City of
Gresham, City of
Portland, City of Bureau of Environmental Services
Water Environment Services

Pennsylvania
Hemlock Municipal Sewer Cooperative (HMSC)
Philadelphia, City of
University Area Joint Authority

South Carolina
Charleston Water System
Mount Pleasant Waterworks & Sewer Commission
Spartanburg Water

Tennessee
Cleveland Utilities
Murfreesboro Water & Sewer Department
Nashville Metro Water Services

Texas
Austin, City of
Dallas Water Utilities
Denton, City of
El Paso Water Utilities

Fort Worth, City of
Houston, City of
San Antonio Water System
Trinity River Authority

Utah

Salt Lake City Corporation

Virginia

Alexandria Sanitation Authority
Arlington, County of
Fairfax County
Hampton Roads Sanitation District
Hanover, County of
Henrico, County of
Hopewell Regional Wastewater Treatment Facility
Loudoun Water
Prince William County Service Authority
Richmond, City of
Rivanna Water & Sewer Authority

Washington

Everett, City of
King County Department of Natural Resources
Seattle Public Utilities
Sunnyside, Port of
Yakima, City of

Wisconsin

Green Bay Metro Sewerage District
Kenosha Water Utility
Madison Metropolitan Sewerage District
Milwaukee Metropolitan Sewerage District
Racine, City of
Sheboygan Regional Wastewater Treatment
Wausau Water Works

Australia

ACTEW (Ecowise)
South Australian Water Corporation
South East Water Limited
Sydney Water Corporation
Water Corporation of Western Australia

Canada

Edmonton, City of/Edmonton Waste Management Centre of Excellence
Lethbridge, City of
Regina, City of, Saskatchewan
Toronto, City of, Ontario
Winnipeg, City of, Manitoba

New Zealand

Watercare Services Limited

United Kingdom

Anglian Water Services, Ltd.

STORMWATER UTILITY

California

Fresno Metropolitan Flood Control District
Los Angeles, City of, Department of Public Works
Monterey, City of
San Francisco, City & County of
Santa Rosa, City of
Sunnyvale, City of

Colorado

Aurora, City of
Boulder, City of

Florida

Orlando, City of

Iowa

Cedar Rapids Wastewater Facility
Des Moines, City of

Kansas

Overland Park, City of

Kentucky

Louisville & Jefferson County Metropolitan Sewer District

Maine

Portland Water District

North Carolina

Charlotte, City of, Stormwater Services

Pennsylvania

Philadelphia, City of

Tennessee

Chattanooga Stormwater Management

Texas

Harris County Flood Control District, Texas

Washington

Bellevue Utilities Department
Seattle Public Utilities

STATE

Connecticut Department of Environmental Protection
Kansas Department of Health & Environment
New England Interstate Water Pollution Control Commission (NEIWPCC)
Ohio River Valley Sanitation Commission
Urban Drainage & Flood Control District, CO

CORPORATE

ADS Environmental Services
Advanced Data Mining International
AECOM
Alan Plummer & Associates
Alpine Technology Inc.
Aqua-Aerobic Systems Inc.
Aquateam–Norwegian Water Technology Centre A/S
ARCADIS
Associated Engineering
Bernardin Lochmueller & Associates
Black & Veatch
Blue Water Technologies, Inc.
Brown & Caldwell
Burgess & Niple, Ltd.
Burns & McDonnell
CABE Associates Inc.
The Cadmus Group
Camp Dresser & McKee Inc.
Carollo Engineers Inc.
Carpenter Environmental Associates Inc.
CET Engineering Services
CH2M HILL
CRA Infrastructure & Engineering
CONTECH Stormwater Solutions
D&B/Guarino Engineers, LLC
Damon S. Williams Associates, LLC
Ecovation
EMA Inc.
Environmental Operating Solutions, Inc.
Environ/The ADVENT Group, Inc.
Fay, Spofford, & Thorndike Inc.
Freese & Nichols, Inc.
ftn Associates Inc.
Gannett Fleming Inc.
Garden & Associates, Ltd.
Geosyntec Consultants
GHD
Greeley and Hansen LLC
Hazen & Sawyer, P.C.
HDR Engineering Inc.
HNTB Corporation
Hydromantis Inc.
HydroQual Inc.
Infilco Degremont Inc.
Jacques Whitford NAWE, Inc.
Jason Consultants LLC Inc.
Jordan, Jones, & Goulding Inc.
KCI Technologies Inc.
Kelly & Weaver, P.C.
Kennedy/Jenks Consultants
KMK Consultants
Larry Walker Associates
Limno-Tech Inc.
The Low Impact Development Center Inc.
Malcolm Pirnie Inc.
Material Matters
McKim & Creed
MPR Engineering Corporation, Inc.
MWH
NTL Alaska, Inc.
O'Brien & Gere Engineers Inc.
Odor & Corrosion Technology Consultants Inc.
Paradigm Environmental Technologies, Inc.
Parametrix Inc.
Parsons
Post, Buckley, Schuh & Jernigan
Praxair, Inc.
Ring Industrial Group
RMC Water & Environment
Ross & Associates Ltd.
SAIC
Siemens Water Technologies
The Soap & Detergent Association
Smith & Loveless, Inc.
Southeast Environmental Engineering, LLC
Stearns & Wheler, LLC
Stone Environmental Inc.
Stratus Consulting Inc.
Synagro Technologies Inc.
Tetra Tech Inc.
Trojan Technologies Inc.
Trussell Technologies, Inc.
Uni-Bell PVC Pipe Association
URS Corporation
Westin Engineering Inc.
Wright Water Engineers
Zoeller Pump Company

INDUSTRY

American Electric Power
American Water
Anglian Water Services, Ltd.
Chevron Energy Technology
The Coca-Cola Company
Dow Chemical Company
DuPont Company
Eastman Chemical Company
Eli Lilly & Company
Johnson & Johnson
Merck & Company Inc.
Procter & Gamble Company
Suez Environnment
United Utilities North West (UUNW)
United Water Services LLC

Note: List as of 2/21/09

WERF Board of Directors

WERF Research Council